A Guide to Hands-on MEMS Design and Prototyping

Whether you are a student taking an introductory MEMS course or a practicing engineer who needs to get up-to-speed quickly on MEMS design, this practical guide provides the hands-on experience needed to design, fabricate, and test MEMS devices. You'll learn how to use foundry multiproject fabrication processes for low-cost MEMS projects, as well as computer-aided design tools (layout, modeling) that can be used for the design of MEMS devices. Numerous design examples are described and analyzed, from fields including micromechanics, electrostatics, optical MEMS, thermal MEMS, and fluidic MEMS. There is also a chapter on packaging and testing MEMS devices, as well as exercises and design challenges at the end of every chapter. Solutions to the design challenge problems are provided online.

JOEL A. KUBBY is a Professor of Electrical Engineering in the Baskin School of Engineering at the University of California, Santa Cruz. Prior to this, he was an Area Manager with the Xerox Wilson Center for Research and Technology, and a Member of the Technical Staff at the Webster Research Center in Rochester, New York. He has led a six-company industrial research consortium under the National Institute of Standards and Technology's Advanced Technology Program (ATP) to develop a new process for optical MEMS, and he has more than 80 patents.

A Guide to Hands-on MEMS Design and Prototyping

JOEL A. KUBBY
University of California, Santa Cruz

CAMBRIDGE
UNIVERSITY PRESS

Shaftesbury Road, Cambridge CB2 8EA, United Kingdom

One Liberty Plaza, 20th Floor, New York, NY 10006, USA

477 Williamstown Road, Port Melbourne, VIC 3207, Australia

314–321, 3rd Floor, Plot 3, Splendor Forum, Jasola District Centre, New Delhi – 110025, India

103 Penang Road, #05–06/07, Visioncrest Commercial, Singapore 238467

Cambridge University Press is part of Cambridge University Press & Assessment,
a department of the University of Cambridge.

We share the University's mission to contribute to society through the pursuit of
education, learning and research at the highest international levels of excellence.

www.cambridge.org
Information on this title: www.cambridge.org/9780521889254

© Cambridge University Press & Assessment 2011

This publication is in copyright. Subject to statutory exception and to the provisions
of relevant collective licensing agreements, no reproduction of any part may take
place without the written permission of Cambridge University Press & Assessment.

First published 2011

A catalogue record for this publication is available from the British Library

Library of Congress Cataloging-in-Publication data
Kubby, Joel A.
A guide to hands-on MEMS design and prototyping / Joel Kubby.
p. cm.
Includes bibliographical references and index.
ISBN 978-0-521-88925-4 (Hardback) – ISBN 978-1-107-64579-0 (Paperback)
1. Microelectromechanical systems. I. Title.
TK7875.K83 2011
621.381–dc22
2011004247

ISBN 978-0-521-88925-4 Hardback
ISBN 978-1-107-64579-0 Paperback

Cambridge University Press & Assessment has no responsibility for the persistence
or accuracy of URLs for external or third-party internet websites referred to in this
publication and does not guarantee that any content on such websites is, or will
remain, accurate or appropriate.

Contents

Preface		*page* ix
1	**Introduction**	1
	1.1 Overview of MEMS fabrication	1
	1.2 Shared wafer processes	6
	1.2.1 Multiproject wafer processes	6
	1.3 Design rules	21
	1.4 Layout	23
	Problems	29
	References	32
2	**Micromechanics**	34
	2.1 Springs	34
	2.1.1 Springs connected in parallel	36
	2.1.2 Springs connected in series	36
	2.2 Buckling	37
	2.3 Poisson's ratio	38
	2.4 Shear stress and strain	40
	2.5 Beams in other situations	41
	2.6 Torsion	43
	2.7 Membranes	44
	2.8 Test structures	45
	2.9 Dampening	48
	2.10 Accelerometer	49
	2.10.1 Cantilever beam	49
	2.10.2 Crash sensor	50
	2.11 Pressure sensor	52

		Problems	53
		References	57
3	**Electrostatic actuation**	58
	3.1	Mechanical restoring force	61
	3.2	Comb-drive resonator	65
	3.3	Cantilever beam resonator	67
	3.4	Fixed-fixed beam resonator	68
		Problems	68
		References	73
4	**Optical MEMS**	74
	4.1	Reflecting cantilever beam optical modulator	74
	4.2	Single-axis torsional mirror	77
	4.3	Dual-axis torsional mirror: Lucent lambda router optical switch	81
	4.4	Fabry-Perot interferometer in the PolyMUMPS process	86
	4.5	Obtaining flatness in optical MEMS devices	93
		Problems	95
		References	96
5	**Thermal MEMS**	98
	5.1	Thermal actuator	101
	5.2	Heatuator	103
	5.3	Thermal bimorph	109
	5.4	Bolometer	112
	5.5	Thermal inkjet	114
	5.6	Thermal damage limits in thermally actuated MEMS	114
		Problems	115
		References	117
6	**Fluidic MEMS**	118
	6.1	Equations of motion	118
	6.2	Microfluidics	119
		6.2.1 Reynolds number	121
		6.2.2 Surface tension	123
		6.2.3 Contact angle	125
		6.2.4 Capillary rise	126
	6.3	Inkjet	127

	Problems	133
	References	134
7	**Package and test**	135
	7.1 Release	135
	7.2 Test equipment	137
	7.3 Mechanical testing	139
	7.4 Electrical testing	139
	7.5 Optical characterization	141
	References	143
8	**From prototype to product: MEMS deformable mirrors for adaptive optics**	144
	References	156
	Index	158

Colour plates follow page 52

Preface

The idea for this book came from a textbook I used in graduate school at Cornell University titled *Introduction to VLSI Systems*, by Carver Mead and Lynn Conway. That textbook, in combination with the MOS Implementation System (MOSIS) Service for integrated circuit prototyping and small volume production, enabled a "hands-on" learning experience that was instrumental in training a new generation of practitioners in very large scale integrated (VLSI) circuit design, layout, and prototyping. This approach democratized VLSI chip design and fabrication by reducing the cost of VLSI circuit prototyping and shortened the turnaround time from years to months, enabling students to design, lay out, and submit chips for fabrication in engineering classes. By providing generic design rules the students did not have to worry about the details for the specific fabrication process. By aggregating small projects into multiproject chips (MPCs) and MPCs into multichip wafers (MCWs), the fabrication cost was decreased by orders of magnitude.

In order to train a new generation of practitioners in MEMS design and prototyping, it is important for students to get a similar hands-on experience. However, hands-on courses on the design, prototyping, and testing of microelectromechanical systems (MEMS) has largely been restricted to universities with cleanroom facilities for semiconductor fabrication. The number of universities with a cleanroom is limited and cleanrooms are expensive to maintain. In addition, they provide too much design freedom that can promulgate the mindset of needing a new MEMS fabrication process for each new MEMS device.

The goal of this text is to guide the student through a MEMS design experience using state-of-the-art computer-aided design tools for layout and modeling, and to submit the design for fabrication in a robust, standard multiproject wafer fabrication process. Many of these processes have mature design rules that enable the separation of design and fabrication challenges. If the rules are followed, the students will get what they have designed. While the use of a standard process leads to limitations in the devices that can be fabricated, since the layers and layer thicknesses are set by the standard process, this approach has been used successfully for the development of a number of commercially successful MEMS devices, some of which are described in the text. By using a robust standard multiproject wafer fabrication process, a working prototype device can be fabricated quickly at low cost. The book also analyzes some standard MEMS designs such as the mechanical test (M-Test) structures that were developed by Professor Stephen Senturia's group at MIT. The M-Test structures are straightforward to design and lay out and provide simple MEMS structures for post-fabrication evaluation and testing that do not require specialized metrology equipment. By simply measuring the pull-in voltages of these structures, which can be done with just a probe station and a high-voltage power supply, validation of the design can be obtained and fundamental information about the prototyping process can be extracted.

The book includes a number of examples from student projects in undergraduate (EE115) and graduate (EE215) MEMS design courses that I have taught in the Department of Electrical Engineering at the University of California Santa Cruz. In the quarter-long course the students learn about microelectromechanical systems in various application domains (mechanics, electrostatics, optical, thermal, and fluidic) and then propose a design of their own to meet a set of specifications that they are provided with. They then design their own part and use modeling and simulation to test their designs. Finally, they lay out their designs using state-of-the-art tools and submit them for fabrication on a class multiproject chip that is then aggregated into a multichip wafer for fabrication using a standard process. I encourage all students using standard multiproject wafer fabrication processes to submit reports on their projects for publication in future editions of this book.

The book ends with a case study of a commercial MEMS device, a deformable mirror for applications in adaptive optics, first prototyped using the PolyMUMPS process, that eventually went on to become a product offered by Boston Micromachines Corporation. This case study demonstrates that there is a "path-to-the-sea" from MEMS prototyping in a standard multiproject wafer fabrication process to commercialization of a robust MEMS product. Once the prototype was designed, fabricated, and tested using a multiproject wafer fabrication process, PolyMUMPS, the standard process was modified as required to meet the particular product specifications. Future editions will include additional case studies of MEMS products that are developed from prototypes fabricated in standard multiproject wafer processes.

In memory of ERK, LJK and LSK, and in appreciation of RSNK-A, who took the brave steps to avoid the fate of the genetic hand she was dealt.

Joel Kubby
University of California, Santa Cruz

1
Introduction

1.1 Overview of MEMS fabrication

Microelectromechanical systems (MEMS) fabrication developed out of the thin-film processes first used for semiconductor fabrication. To understand the unique features of the MEMS fabrication process it is helpful to consider the semiconductor fabrication process.

The semiconductor fabrication process is cyclic. (a) First, a thin film is deposited on the wafer surface using thin-film deposition techniques. (b) A uniform photosensitive polymer (photoresist) is then deposited and (c) exposed to light from a mask that contains the pattern that is desired on the thin film. (d) The photoresist is developed to obtain the desired pattern. (e) The pattern in the photoresist is then transferred to the thin film using an etching technique, and the photoresist is removed. Figure 1.1 shows a cross section of the wafer at each step. This cycle is repeated for each new layer, with some processes requiring as many as 20 to 30 cycles.

The typical thin films that are deposited include semiconductors (e.g., polysilicon), insulators (e.g., silicon nitride), and metals (e.g., aluminum). In addition, some layers are grown (oxide), diffused, or implanted (dopants) rather than deposited using thin-film techniques. A cross section of a complementary metal oxide semiconductor (CMOS) process that includes six levels of metal is shown in Figure 1.2 [1]. A schematic diagram of one of the first MEMS devices, which used semiconductor processing for fabrication, the resonant gate transistor, is shown in Figure 1.3 [2].

The MEMS fabrication cycle, as shown in Figure 1.4, also includes thin-film deposition and patterning, but typically the films are thicker than the films used in microelectronics, the etching is deeper, and the

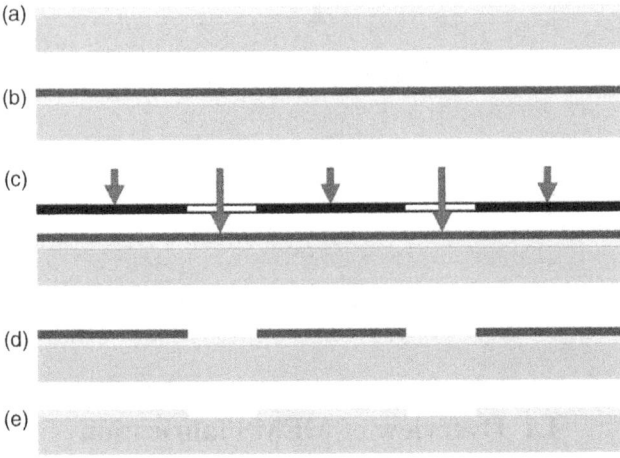

Figure 1.1 The semiconductor fabrication process. (a) Thin-film deposition (yellow), (b) photoresist deposition (blue), (c) photolithography (mask clear and opaque; red arrows), (d) photoresist development, and (e) etching to transfer the pattern in the photoresist into the thin film. See color plate section.

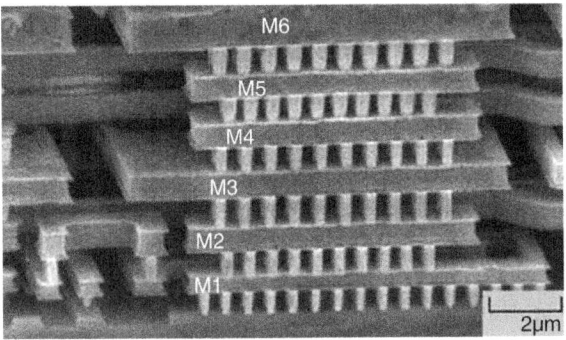

Figure 1.2 A cross-section scanning electron microscope image of a six-level metal backend structure. The insulating layers between the metal layers have been etched away, similar to the sacrificial etch that is used to release structures in the polysilicon surface micromachining process described below. (Reprinted with permission from JOM Journal of the Minerals, Metals and Materials Society.)

photolithography is more challenging because the topography from the patterning is greater [3]. The films and etches can be several microns deep, leading to topography approaching 10 μm after seven or eight layers have been deposited and patterned. Another difference is

1.1 Overview of MEMS fabrication

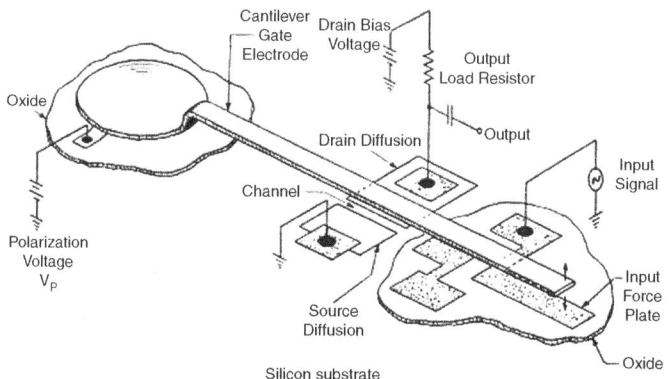

Figure 1.3 The resonant gate field effect transistor, one of the first MEMS devices. A released metal cantilever beam forms the gate electrode over the diffused source–drain channel. The input signal is applied to the input force plate, which causes the cantilever beam to vibrate, modulating the current through the transistor. Maximum vibration occurs at the resonant frequency of the cantilever beam, enabling the device to act as a high-Q electromechanical filter. (Reprinted with permission from IEEE Trans. Electron Devices, *The resonant gate transistor*, H.C. Nathanson, W.E. Newell, R.A. Wickstrom and J.R. Davis Jr., ©1967 IEEE.)

that the MEMS fabrication process ends with the release of mechanical elements such as beams and membranes. Since some elements are released, the mechanical properties of the deposited films must be controlled to avoid distortion of the released elements as they relax under the influence of residual stress and stress-gradients, as shown in Figures 1.5 and 1.6.

Controlling the mechanical properties of thin films can require considerable process development. The mechanical properties are sensitive to how the layers are deposited, what layers they are deposited on, what layers are deposited on them, and the thermal history that the layers are subjected to. As an example, each film has a unique coefficient of thermal expansion (CTE). During a typical thin-film deposition step, the wafer may be heated by hundreds of degrees. When the wafer is cooled the deposited film will have a different rate of dimensional change than the substrate, leading to a buildup of stress.

Typically each new MEMS device requires the development of a new MEMS fabrication process, so controlling the thin-film properties in one process does not necessarily help with control of the thin-film properties

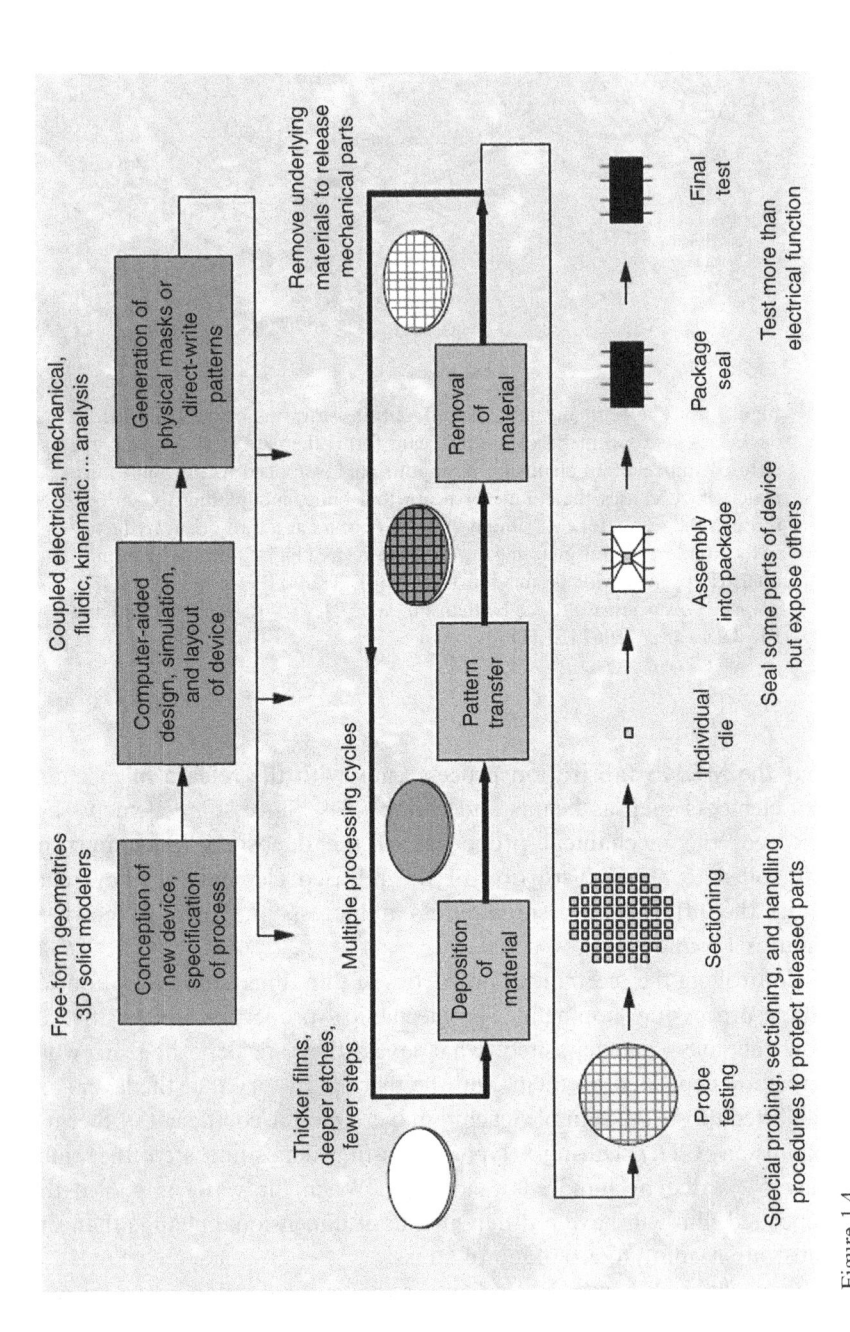

Figure 1.4

1.1 Overview of MEMS fabrication

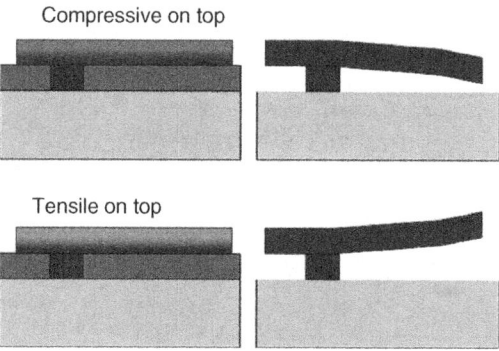

Figure 1.5 Deformation of released structures due to residual stress-gradients. In the top figure the stress is compressive on top, resulting in a downward deflection on release. In the bottom figure the film is tensile on top, resulting in an upward deflection on release. See color plate section.

in a different process. This is in contrast to the control of electrical properties in microelectronics fabrication. Here results from one process, such as the controlled doping of silicon by diffusion or implantation, can be re-applied to a different process. The development of a new MEMS process can take decades and cost millions of dollars, so it can be justified only by market demand for the MEMS device that is eventually manufactured. Here we consider some standard MEMS processes that can be used at the prototyping stage and eventually modified as required to reach the manufacturing stage.

Figure 1.4 The MEMS prototyping cycle. Once a MEMS device has been conceived and a fabrication process specified, it is modeled to the first order analytically and, if required, to a higher order using computer-aided design tools capable of coupled domain analysis (e.g., electrical, mechanical, fluidic, thermal). The design is then finalized and a layout is generated using a layout editor that creates files (GDSII or CIF) that are used to define the physical mask layers. Once the masks have been written they are used to pattern the various layers of the MEMS device in multiple processing cycles. A typical MEMS process might involve eight deposition and patterning cycles. Typically the deposited layers are thicker, and the etches deeper, than those used in integrated circuit fabrication. After all of the layers are deposited and patterned, the wafer is tested, diced (sectioned), and packaged. The sacrificial layers can be removed at either the wafer or the die level. (Reprinted with permission from IEEE Trans. Electron Devices, *MEMS: The systems function revolution*, Karen W. Markus and Kaigham J. Gabriel, ©1967 IEEE.)

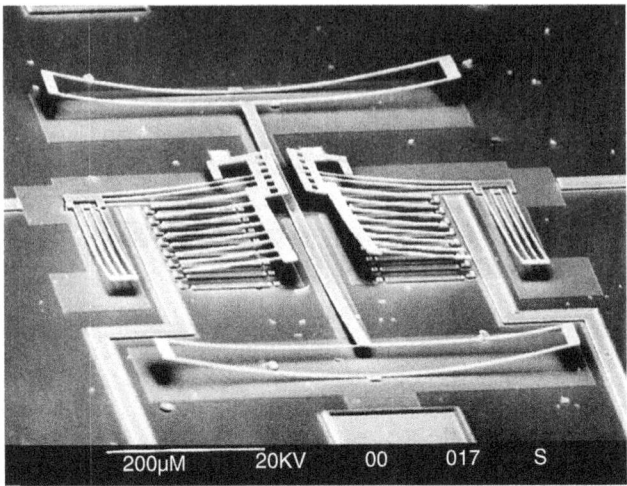

Figure 1.6 Deformation of a released structure due to stress-gradients at MCNC circa 1996. The residual stress is tensile on top of the polysilicon, causing the released structure to bend away from the substrate after release. In general, the polysilicon from MCNC was quite flat, but in this case an anneal of the sacrificial doped oxide (PSG) was left out, resulting in residual stress-gradients. (Reprinted with permission of Kristofer S.J. Pister, Berkeley Sensor and Actuator Center, UC Berkeley.)

1.2 Shared wafer processes

1.2.1 Multiproject wafer processes

There are a number of multiproject wafer (MPW) processes that are available that use bulk and surface micromachining as well as electroplating and wafer bonding. By using a robust MPW process that has stable thin-film parameters and a well-developed set of design rules, the prototyping challenges associated with process development can be avoided so that the focus of the effort can be on the MEMS design. Additional benefits are the decreased costs, because the wafer is shared between many users, and a path-to-the-sea for product manufacturing, because a commercial foundry is involved from the start, providing a low-cost proof of concept to high-volume manufacturing. A price that must be paid is that the process is already defined and cannot be altered. The only layers that are available are the layers that are used in the MPW process. Even the layer thickness cannot be altered. Nonetheless, if a standard MPW process is initially used, the foundry that provides the standard process can usually provide minor changes such as the addition of a new layer or a change to the thickness of an existing layer at the expense of higher fabrication costs.

1.2.1.1 Surface micromachining

There are a number of surface micromachining MPW processes available that include variations in the numbers of layers as well as different layer materials. The earliest MPW surface micromachining process offered was the three-layer polysilicon surface micromachining process initially developed at UC Berkeley by Roger Howe and Richard Muller [4]. This process has been reviewed by Bustillo et al. [5]. A cross-sectional diagram of the layer stack that is currently offered by MEMSCAP in their PolyMUMPS process is shown in Figure 1.7 [6].

In this process, as shown step-by-step in cross section in Figure 1.7, the surface of a 150 mm substrate (n-type, 1–2 Ωcm) is heavily doped with phosphorus to a resistance of 10 $\Omega/\square$ to avoid charge buildup at the substrate–nitride interface when high voltages are applied between the substrate and the subsequent conducting layers. The surface is protected by a blanket low-pressure chemical vapor deposition (LPCVD) of a 0.6 µm thick insulating silicon nitride (90 MPa residual tensile stress). A 0.5 µm thick layer of LPCVD polysilicon (n-type, 30$\Omega/\square$, −25 MPa residual compressive stress) Poly0 is deposited and then patterned. This layer is typically used as a ground plane or counter-electrode and is not released. The next layer is the first oxide layer. It is a 2 µm thick phospho-silicate glass (PSG) that is deposited by LPCVD. This "sacrificial" oxide has a fast etch rate in 49% buffered hydrofluoric acid (HF) that allows for a quick release (2.5 min) of nonsacrificial (i.e., not etched by HF) "structural" layers deposited on top of it if properly spaced etch holes are included in the structural layers, as explained later. The PSG layer is also used as a solid source for phosphorus doping of the polysilicon layers that are in contact with it during the thermal anneal step used to control the level of stress in the polysilicon.

Shallow holes, approximately 0.75 µm deep, which do not go all the way through the 2 µm thick oxide, are defined using the DIMPLE mask. These "dimple" holes are later filled in with Poly1 to form tiny stalactites that hang down from the bottom of Poly1 to limit the amount of surface area contact between Poly1 and Poly0. Surface forces such as van der Waals and capillarity dominate on a microscale, which can lead to surfaces sticking together. This is so prevalent in microelectromechanical

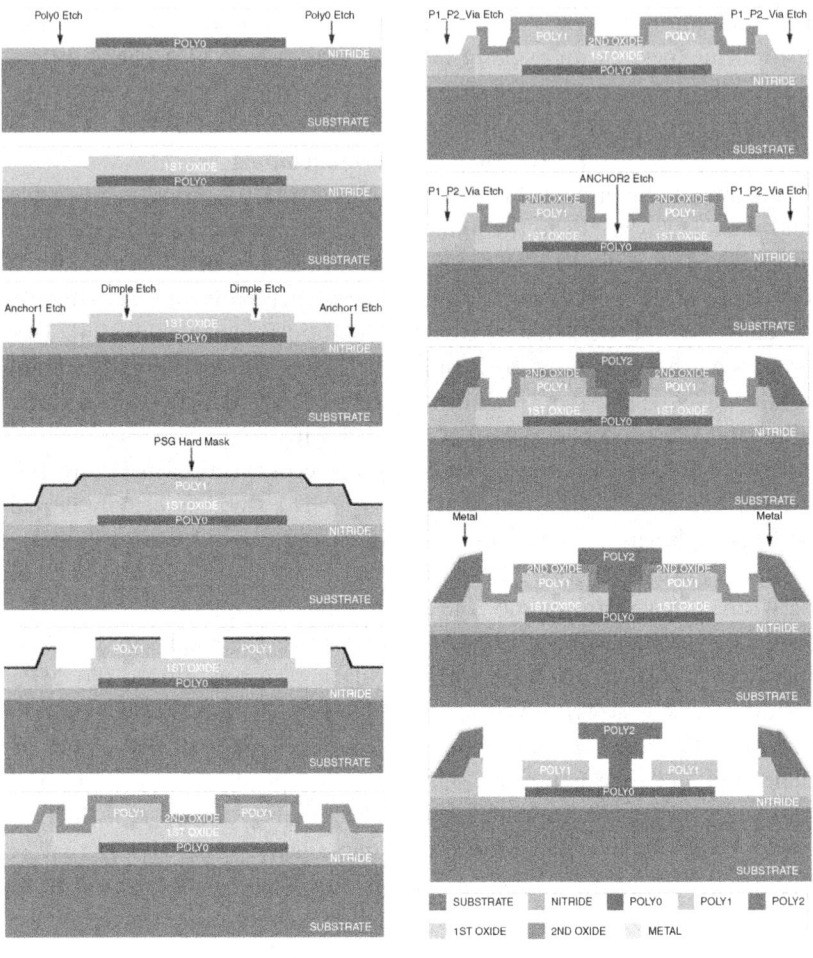

Figure 1.7 PolyMUMPS three-layer polysilicon surface micromachining process offered by MEMSCAP. Polysilicon and oxide layers are deposited and patterned in a cyclic process, with anneal steps of the doped sacrificial oxide between polysilicon depositions. Poly0 is an electrical layer that is not released. Poly1 and Poly2 are structural layers that can be released. The deposition and patterning steps shown here result in a polysilicon wheel defined in Poly1 that is constrained by a hub defined in Poly2. Dimples defined in POLY1 keep the wheel from becoming stuck to the Poly0 layer. (Reprinted with permission from MEMSCAP Inc.) See color plate section.

systems that a new term has been created – "stiction." A recommendation is to include as many dimples as the design allows. One way to make sure dimples are widespread is to include them between every release etch hole, which are spaced every 30 μm, as will be described later.

1.2 Shared wafer processes

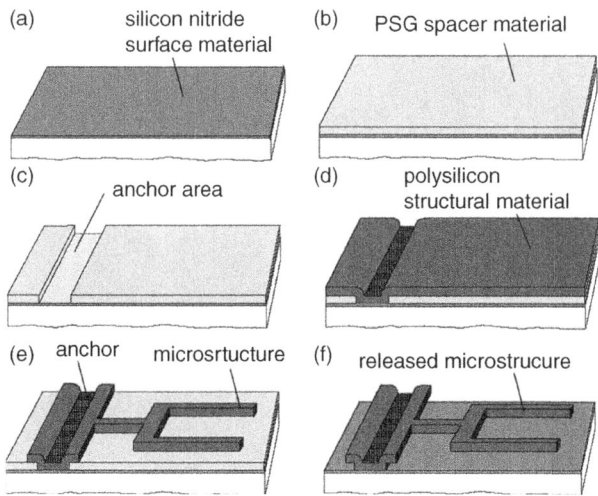

Figure 1.8 Polysilicon surface micromachining. (a) An insulating surface layer such as silicon nitride is deposited on the substrate. (b) A sacrificial spacer layer, such as phospho-silicate glass (PSG), which can be etched quickly, is deposited. (c) A hole is cut through the sacrificial layer to the insulating layer, where the structure is to be anchored to the underlying surface material. (d) The structural layer, in this example polysilicon, is then deposited. The deposition is conformal and fills in the hole that had previously been cut through the sacrificial layer. (e) The structural layer is patterned and then (f) released. (Reprinted with permission from ITC International Test Conference, MEMS Fabrication, Gary K. Fedder, ©1967 IEEE.)

Holes are next cut all the way through the first oxide layer (ANCHOR1) to anchor structures defined in Poly1 to the substrate after the PSG sacrificial spacer layer has been etched. An example is shown in Figure 1.8 [7]. Note that the polysilicon is a conformal coating, so that topographic features, such as the anchor holes, are replicated in the layers deposited on top of them. This replication of topography can give rise to mechanical interferences for features defined in the overlayers.

A common mistake made by those who are first learning to use a surface micromachining process is to leave out anchors. Since you must specify where a hole in this layer is to be placed, it is easy to leave out the anchor hole. If no anchor hole is specified, the polysilicon that is deposited over the oxide will be released during the sacrificial etch and float away. A good practice to avoid this common mistake is to take cross sections of the thin-film stack to make sure that all structures are anchored. An example of a bond pad that was not anchored and was released during the sacrificial etch is shown in Figure 1.9.

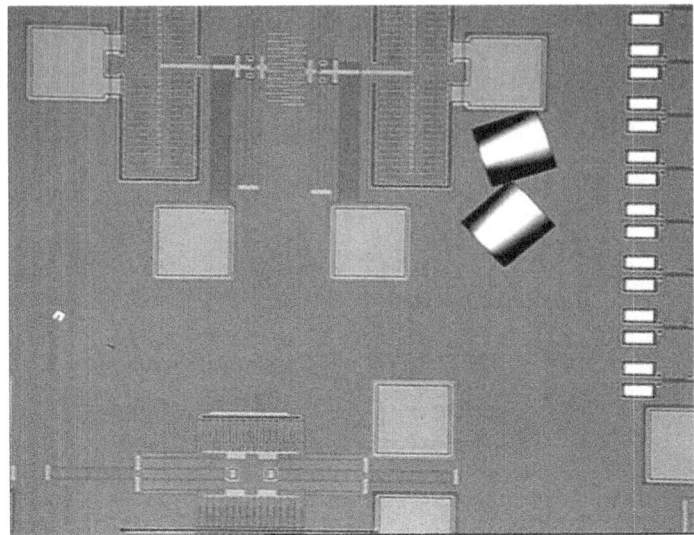

Figure 1.9 Bond pads that were not anchored are released after the sacrificial etch. Not only does this make electrical testing difficult, but the bond pads usually end up in places you do not want them. Your neighbors on a multiproject chip will be particularly annoyed if your parts land in their area! The parts that do not get stuck to the chip are often observed as "floaters" in the etch bath that is used for the sacrificial release.

After the ANCHOR1 holes have been etched, the first layer of structural polysilicon that can be released, Poly1, is deposited and patterned on top of Oxide1. This layer of polysilicon is 2 μm thick. A thin film of PSG (0.2 μm) is deposited on top of Poly1 to act as a solid source for doping the top side of the polysilicon layer, and as a hard mask for subsequent patterning of the layer. The wafer is then annealed at a high temperature (1050°C) for an hour to relieve residual stress and to dope the polysilicon. This high-temperature anneal leads to a low compressive residual stress (-10 MPa) and a resistance of 10 $\Omega/\square$. The polysilicon layer is doped from both the top and the bottom of the thin film simultaneously to minimize stress-gradients through the thickness of the thin film. A stress-gradient is a difference in stress between the top and the bottom layers of a thin film, and it can cause released structures to deform to relieve the stress. A stress-gradient that is compressive on the top causes a released structure to bend down toward the substrate, as shown in Figure 1.5. A thin film that is tensile on top causes the release structure to bend upward, away from the substrate. An example of deformations caused due to stress-gradients in the early days of PolyMUMPS process development is shown in Figure 1.6. In spite of the symmetrical doping from the top and bottom, there will still be a slight stress-gradient that is tensile on top and will cause cantilevers to bow upward on release. The amount of bow will depend on the thickness of the layer and the length of the fixed-free beam.

In addition to defining features in the Poly1 layer where Poly1 will remain, etch holes can be defined in the Poly1 layer using the HOLE1 mask level. These etch holes are required to release Poly1 structures that are more than 30 μm wide, and they should be separated by no more than 30 μm to ensure proper release. If additional layers of polysilicon or gold are deposited on top of Poly1 features, the overlayers will also require etch holes to expose the etch holes defined in Poly1. HOLE2 is used to create etch holes in Poly2 and must surround the HOLE1 features by 2 μm, as described in the PolyMUMPS design rules [8]. HOLEM is used to create etch holes in the gold metal layer.

Following the high-temperature anneal of Poly1 and its subsequent patterning, another sacrificial PSG layer, Oxide2, is deposited that is 0.75 μm thick. This layer allows the release of structures that are defined in the following polysilicon layer, Poly2. Two different anchors are defined in this layer. To anchor structures defined in Poly2 to the substrate, a hole through both oxide layers is patterned using ANCHOR2. Since both layers of oxide are patterned at the same time, misalignments between different mask layers can be avoided. If a feature defined in Poly2 is to be anchored and electrically connected to Poly1 instead of the substrate, POLY1_POLY2_VIA is used.

A common mistake is to use a combination of POLY1_POLY2_VIA and ANCHOR1 to anchor features defined in Poly2 to the substrate, instead of using ANCHOR2. These two approaches for anchoring Poly2 features to the substrate are not equivalent, because Poly0 or nitride can be exposed by the hole defined for ANCHOR1 and etched during subsequent patterning steps since they are unprotected by the cover layer of Poly1 required by the design rules for the ANCHOR1 layer.

After patterning the Oxide2 layer with anchors and vias, the second layer of released structural polysilicon is deposited, annealed, and patterned following the procedures used for Poly1. Poly2 is 1.5 μm thick and has a compressive residual stress of -10 MPa and a resistance of 20 Ω/□. A common use for this structural layer is to provide constraints for features defined in Poly1, such as a hub that constrains a rotor in an electrostatic motor [9], [10] or a staple that constrains a pin in a hinge [11], [12]. It is also possible to make a

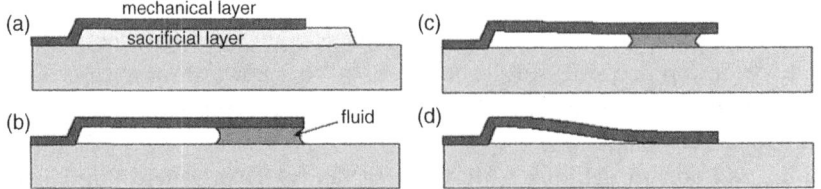

Figure 1.10 Stiction that can occur during the sacrificial release etch. (a) Before sacrificial etching, the sacrificial oxide is below the mechanical layer. (b) After the chip is removed from the etch bath it begins to dry and the remaining fluid forms a bridge between the substrate and the mechanical layer. (c) Capillary forces from the meniscus of the fluid exert a downward force on the cantilever and cause it to come into contact with the substrate. (d) The surface forces, such as Van der Waals attraction, that dominate at the microscale cause the cantilever to become stuck to the substrate. (Reprinted with permission from IOP Publishing Ltd.) [15].

polysilicon layer that is 3.5 µm thick by stacking the Poly2 on top of Poly1. An even thicker, and thus stiffer, structural layer can be formed by trapping the Oxide2 between Poly1 and Poly2 so that it is not etched during the sacrificial release step. This results in a Poly1/Oxide2/Poly2 stack that is 4.25 µm thick.

The final layer is a 0.5 µm thick gold metal layer on top of Poly2 for wires, bond pads, bimorphs, and potentially as an optically reflective surface. The gold is deposited on top of a thin (20 nm) chrome layer to promote adhesion. It is not possible to deposit gold on top of the Poly1 layer. For a flat mirror to be formed, the stress-induced curvature from the metallization should be comprehended in the design [13].

Once the fabrication process has been completed the parts can be released in a sacrificial etch of the exposed oxide layers. It is possible to use a wet buffered HF release or a dry vapor HF release. The dry release helps to avoid the stiction problems that can arise in a wet release, as shown schematically in Figure 1.10. Stiction problems in a wet HF release can be avoided by using critical point drying where the liquid-gas phase during drying is eliminated, and thus also eliminating the formation of a meniscus and the resultant capillary forces that can cause released parts to come into contact with each other and stick together [14], [15]. It is also possible to use self-assembled monolayers that make the resultant surface hydrophobic, which will also eliminate the formation of a meniscus. In all cases dimples help to minimize stiction both during and after release.

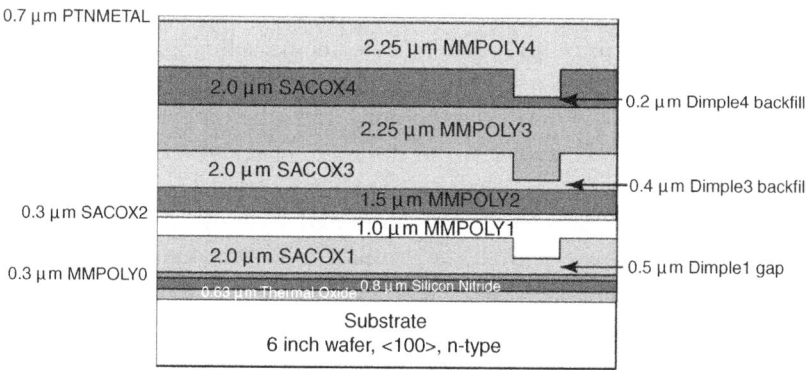

Figure 1.11 SUMMiT V cross section. The SUMMiT V process features five layers of polysilicon, four of which can be released. The upper layers of sacrificial oxide are polished, removing the topography that develops during the conformal polysilicon depositions. (Courtesy of Sandia National Laboratories, SUMMiT(TM) Technologies, www.mems.sandia.gov.) See color plate section.

1.2.1.2 Sandia Ultra-planar Multilevel MEMS Technology V

A polysilicon surface micromachining process with five layers of poly is the Sandia Ultra-planar Multi-level MEMS Technology V (SUMMiT V) process offered by Sandia National Laboratory. A cross section of the layers in this process is shown in Figure 1.11.

Some of the features of this process are that there are four layers of released micromechanical (MM) polysilicon (MMPOLY1–4), enabling the fabrication of more complex structures than the two released polysilcon layers in the PolyMUMPS process, 1 μm design rules, and chemical mechanical polishing (CMP) of the oxide directly beneath the upper two levels of the mechanical polysilicon. The polishing of the upper layers of oxide eliminates the conformal topography that builds up in the PolyMUMPs process and enables high-resolution features to be defined in spite of the thick underlying material stack that would otherwise cause depth of focus problems in photolithography.

This process has been used in early prototyping of MEMS mirror arrays for applications in adaptive optics, optical switching, and deformable grating arrays [16], [17]. The SUMMiT IV process was transferred from Sandia National Laboratories to Fairchild Semiconductor International but has since been discontinued. Fairchild was the only company that offered this technology for foundry manufacturing.

1.2.1.3 SOIMUMPS

The SOIMUMPS process is based on bulk micromachining of a silicon on insulator (SOI) wafer using four mask levels. It was originally developed for the fabrication of MEMS variable optical attenuators (VOAs) based on the use of a thermal actuator to control an optical shutter [18]. A cross-sectional diagram of an SOI wafer is shown in Figure 1.12.

The use of bulk micromachining of an SOI wafer offers a number of advantages over the polysilicon surface micromachining processes described previously. First, the use of bulk micromachining enables the definition of thick structures. The availability of thick structures is useful for optical devices in which the silicon is metallized to form a mirror surface. Metallization of the thin polysilicon layers in the PolyMUMPS process can cause the surfaces to deform due to the stress in the metal layer [8]. The structures are fabricated in the device layer of the SOI wafer, with a choice of device layer thicknesses of 10 ± 1 μm or 25 ± 1 μm. Since the minimum design rule in patterning the SOI layer is 2 μm, features with aspect ratios of 5:1 to 10:1 are possible. Second, the device layer of the SOI wafer is single crystal silicon with excellent and well-controlled electrical and mechanical properties [19]. Third, the 1 μm thick buried oxide layer in the SOI wafer provides a built-in etch stop for bulk micromachining from the front and back sides of the wafer, simplifying the number of processing steps. A cross-sectional view of the patterned wafer is shown in Figure 1.13.

The process uses four mask levels (PAD METAL, SOI, TRENCH, and BLANKET METAL) that are used to pattern fine metal features on the device layer, holes through the device layer, holes through the substrate, and shadow-masked metal features on the device and substrate layers. The process starts off with a 4 in. SOI wafer (substrate 1–10 Ω-cm, n-type, device layer 1–10 Ω-cm) that is heavily phosphorus doped at the surface (15–25 $\Omega/\square$) by solid source diffusion from a PSG layer during a 1 h anneal at 1050°C in argon. The PSG is then stripped in a wet etch.

The silicon device layer is metallized with gold (500 nm Au/20 nm Cr) and patterned with the PAD METAL mask using a photolithographic lift-off process that is capable of defining 3 μm lines and spaces with a 3 μm alignment tolerance. This metal layer is exposed to high temperatures during the subsequent process steps, so it does not provide an optical quality surface for mirrors like the second metallization that is patterned with the BLANKET METAL mask. Any metal features that are defined in the first metal deposition will be in electrical contact unless they are separated by a trench etched in the device layer since the surface of the device layer is heavily doped with phosphorus.

1.2 Shared wafer processes

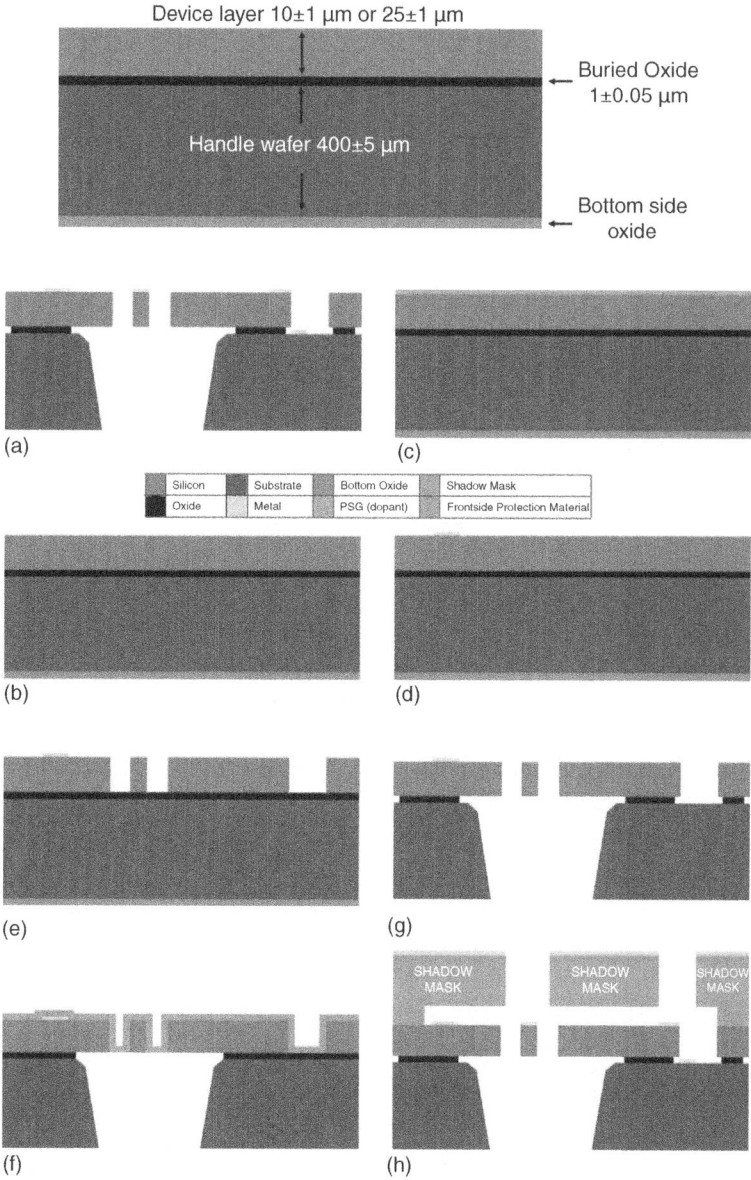

Figure 1.12 Silicon on insulator (SOI) wafer that is used in the SOIMUMPS process. The device layer can be 10 ± 1 μm or 25 ± 1 μm thick. The handle wafer is 400 ± 5 μm thick and the buried oxide layer is 1 ± 0.05 μm thick. (Reprinted with permission from MEMSCAP Inc.) See color plate section.

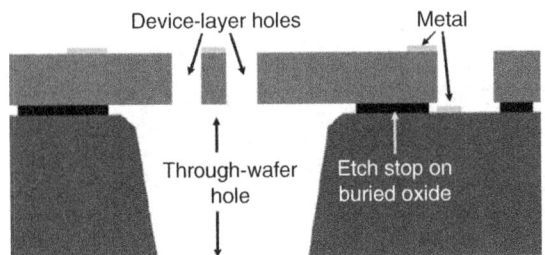

Figure 1.13 Patterned SOIMUMPS wafer. Through-wafer etches are performed from the front side of the wafer 10 or 25 μm deep to form device layer holes, and from the back side 400 μm deep to form through-wafer holes. Both etches stop on the buried oxide. (Reprinted with permission from MEMSCAP Inc.) See color plate section.

The device layer holes are patterned photolithographically using the second-level SOI mask, and then the pattern is transferred into the device layer using a deep reactive ion etch (DRIE) that has been developed to avoid lateral spreading of the etch (footing) when the etch stops on the buried oxide layer. The front side of the wafer is then protected with a polymer-based resist, and the wafers are flipped over for patterning the backside oxide using the TRENCH mask. The oxide is patterned with reactive ion etching (RIE) followed by a DRIE to form the through-wafer holes, stopping on the buried oxide layer. The buried oxide layer exposed in the through-wafer etch is then removed in a wet etch. The front-side protective coating is then removed in a dry etch, and the remaining oxide is removed in a dry vapor phase HF etch to minimize stiction.

A second layer of gold metallization (600 nm Au/50 nm Cr) is then deposited using the BLANKETMETAL as a shadow mask for patterning. The use of a shadow mask, where metal is evaporated through holes in the mask, avoids problems with the large topography created by the SOI etch. Any metal features on the device layer will not be in contact with metal features deposited on the substrate because the oxide etch results in an undercutting of the buried oxide. The shadow mask evaporation that is used to deposit the second metal is a line-of-site deposition that will not lead to a continuous electrical conductor between metal on the device layer and metal on the substrate layer. The BLANKETMETAL mask has a step etched into it that is brought in contact with the SOI wafer in an exclusion zone outside the 9 mm × 9 mm drawing area available to the user on the chip.

Some examples of optical devices that have been fabricated using SOIMUMPS process are shown in Figure 1.14.

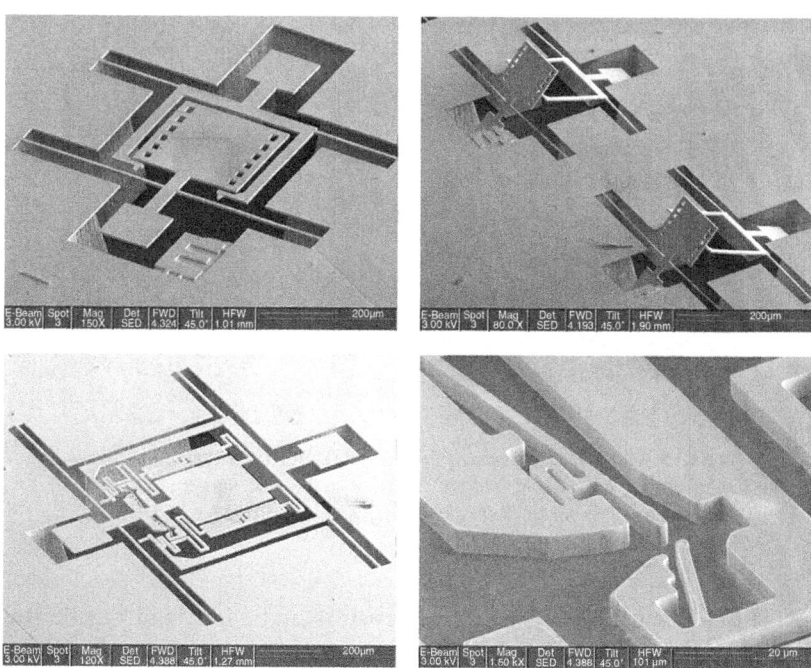

Figure 1.14 Examples of optical MEMS devices that have been fabricated in the SOIMUMPS process. When the square pad attached to the mirror is pushed down with a microprobe, the mirror is lifted up. The torsion hinges allow the mirror to move up while at the same time being attached to the silicon layer. When the mirror is in the upright position, the square pad attached to the locking mechanism can be lifted and its teeth engaged into the holes in the mirror, locking the mirror in its upright position. (Reprinted with permission from Prof. Ash Parameswaran, School of Engineering Science, Simon Fraser University, Burnaby, BC, Canada.)

1.2.1.4 MetalMUMPS

The MetalMUMPS process integrates electroplating and surface and bulk micromachining to form tall (18–22 μm) structures in nickel. The minimum feature size in the nickel layer is 5 μm, so high-aspect-ratio structures (4:1) can be fabricated. It was first developed for fabricating microrelay products, but it can also be used for fabricating radio frequency (RF), magnetic, and microfluidic devices. It is similar in principle to the LIGA process, a German acronym for lithography (Lithographie), Electroplating (Galvanoformung), and Molding (Abformung), but it also incorporates elements of surface micromachining (polysilicon, nitride, sacrificial PSG oxide), and bulk micromachining (KOH anisotropic etching). The electrodeposited nickel is used in the fabrication of

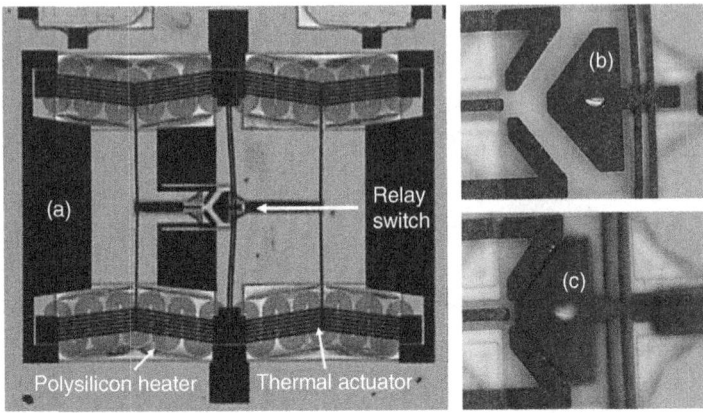

Figure 1.15 Microrelay fabricated in the MetalMUMPS process. (a) Overall relay including polysilicon heaters, thermal actuators, and a relay switch. (b) Relay switch open. (c) Relay switch closed. (Reprinted with permission from MEMSCAP Inc.)

structural elements, polysilicon for resistors and wires, and nitride for electrical insulation. The nickel sidewalls can be further coated with a thin gold layer to improve the electrical conductivity for improved electrical contact in relays. Structural elements can be released by either using sacrificial oxide or undercutting with trenches formed in a bulk KOH etch. An example of a microrelay formed in the MetalMUMPS process is shown in Figure 1.15. A cross section showing all of the layers used to fabricate the relay is shown in Figure 1.16.

The MetalMUMPs process flow starts off with a high-resistivity (>4000 Ω-cm) n-type (100) silicon wafer. A 2 µm thick isolation thermal oxide is grown on the wafer surface that provides electrical insulation from the substrate, followed by a 0.5 µm thick PSG sacrificial oxide (Oxide1) that can be used for releasing structures defined in the Nitride1 layer and for defining regions where the silicon trench will later be etched. The resulting thin-film stack is shown in cross section in Figure 1.17(a). Oxide1 is then patterned with the OXIDE1 mask and wet etched, as shown in Figure 1.17(c). Next, a 0.35 µm thick layer of low-stress nitride (Nitride1) is deposited. The nitride layers provide a protective encapsulation for the polysilicon. The nitride pattern also defines a protective layer on the substrate that determines where Si trench etching occurs. Third, a released and patterned nitride area may also be used to provide a mechanical linkage between released metal structures that must be isolated electrically. The nitride deposition is followed by a 0.70 µm

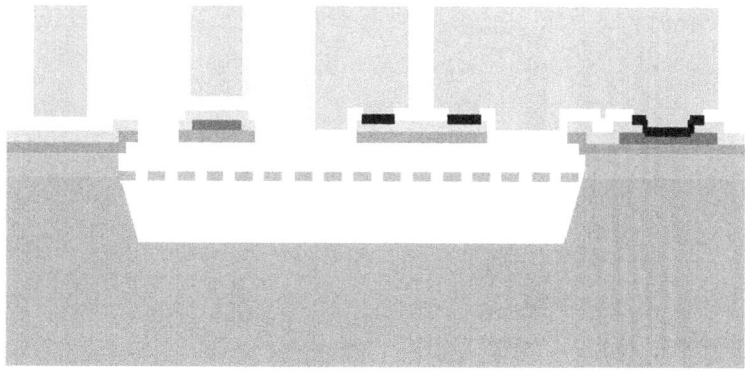

Figure 1.16 Cross section through the layer stack used to fabricate the microrelay. (Reprinted with permission from MEMSCAP Inc.) See color plate section.

thick layer of polysilicon (Poly1), as shown in Figure 1.17(d). The polysilicon is doped using ion implantation followed by a thermal anneal with a final nominal resistance of 22 $\Omega/\square$. This Poly1 layer is then patterned with the POLY mask, as shown in Figure 1.17(d). This layer can be used to form resistor elements or mechanical structures, or for electrical crossover routing.

A second 0.35 μm thick, low-stress nitride layer (Nitride2) is deposited (Figure 1.17[e]), and the combination of Nitride1 and Nitride2 layers is lithographically patterned with the third mask level, NITRHOLE, and etched, as shown in Figure 1.17(f). The second sacrificial layer of PSG oxide (1.1 μm), Oxide2, is deposited (Figure 1.17[g]) and patterned (Figure 1.17[h]) with the fourth mask level, METANCH, and is then wet chemically etched. The Oxide2 patterning step also provides the pattern for the metal structure anchors. A liftoff process is used to provide thin layers of Cr (10 nm) and Pt (25 nm) (Anchor Metal) only in the bottom of the Oxide2 anchors (Figure 1.17[h]).

A plating base consisting of 500 nm Cu and 50 nm of Ti is deposited (not shown in Figure 1.17). The plating base layer provides electrical continuity across the wafer for the subsequent metal electroplating step. The wafers are then coated with a thick layer of photoresist and patterned with the fifth mask level (METAL), forming the stencil that will be filled with electroplated nickel (metal), as shown in Figure 1.17(i).

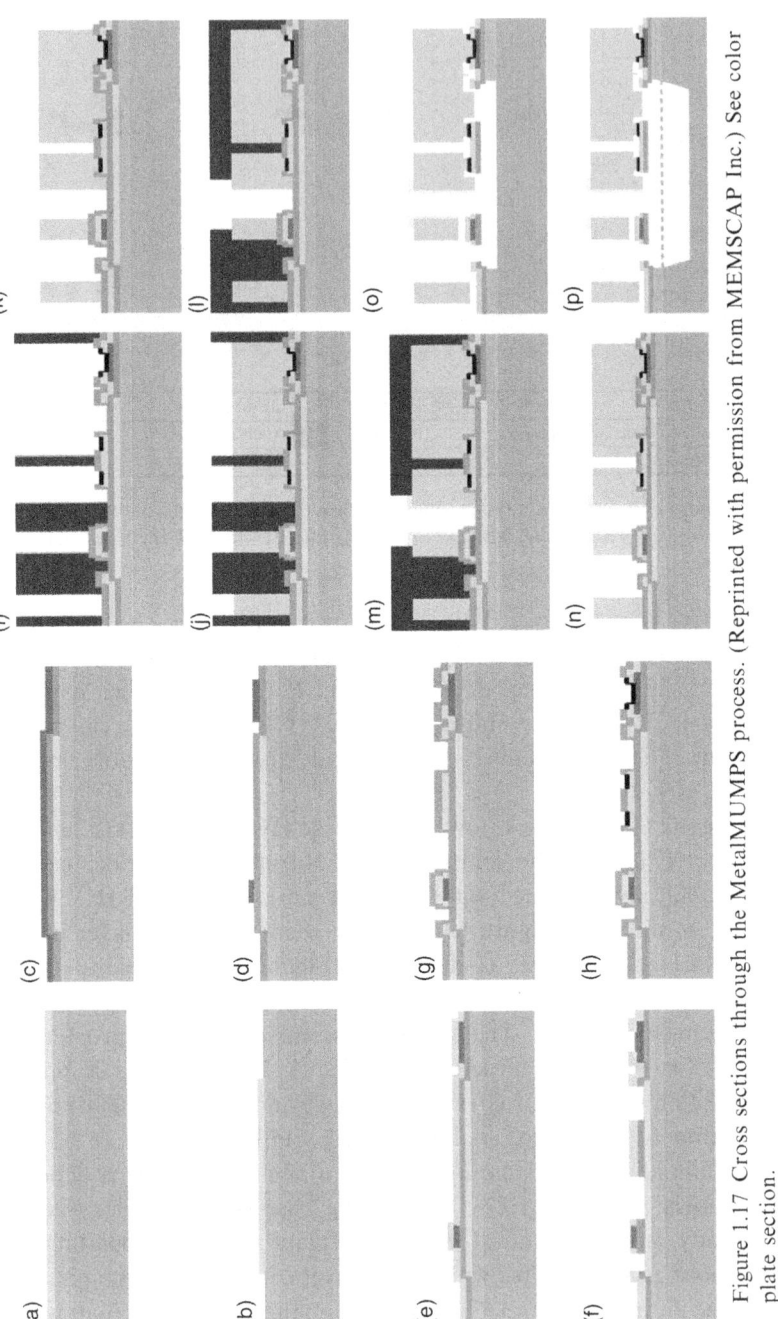

Figure 1.17 Cross sections through the MetalMUMPS process. (Reprinted with permission from MEMSCAP Inc.) See color plate section.

Nickel (8 μΩ cm) is then electroplated in the photoresist stencil up to a nominal thickness of 20 μm (Figure 1.17[j]) and then capped with a 0.5 μm gold layer that forms a pad material suitable for wire bonding (not shown). The photoresist stencil is then removed (Figure 1.17[k]), and a new plating stencil for gold overplating the exposed sidewalls is applied and patterned with the GOLDOVP (Figure 1.17[l]). First an enlarged version of this mask is applied and used to remove the plating base in regions where the sidewall metal is to be applied. The plating base is chemically etched and then the enlarged version of the GOLDOVP mask is removed, and a second layer of photoresist is patterned with an unenlarged GOLDOVP mask.

A layer of gold, 1–3 μm thick, is then plated onto the exposed sidewalls (Figure 1.17[m]). The GOLDOVP resist stencil is removed (Figure 1.17[n]), and the plating base is removed. A 49% HF solution is used to remove the sacrificial oxide layers (Oxide1 and Oxide2), and the isolation oxide over the trench areas (Figure 1.17[o]). Finally, an anisotropic KOH etch is used to form a 25 μm deep trench in the areas defined by the OXIDE1 and NITRHOLE mask layers (Figure 1.17[p]).

1.3 Design rules

Design rules are a set of guidelines that, if followed, ensure that the MEMS fabrication process is able to fabricate what the designer intends. The rules comprehend the processing steps and interactions between steps that, if followed, will allow the user to get what they want. Examples of the issues that are comprehended by design rules include the limits of photolithography such as minimum feature sizes, the alignment between different layers, and pattern transfer fidelity on a surface with a builtup topography. The design rules also comprehend the limits of fabrication, such as the etch selectivity of different thin film layers that are exposed during a pattern transfer step, process latitude for overetching to ensure the complete removal of a film, and how far a structure will be undercut during a sacrificial etch step with a prescribed release etch time to ensure release. Some examples of what can happen if the design rules are not followed are illustrated in Figure 1.18.

In the upper part of Figure 1.18, the first oxide layer in the PolyMUMPS process has been patterned with the ANCHOR1 mask level, with the etch stopping on the nitride layer on the left and on the Poly0 later on the right. A polysilicon layer, Poly1, is then deposited and

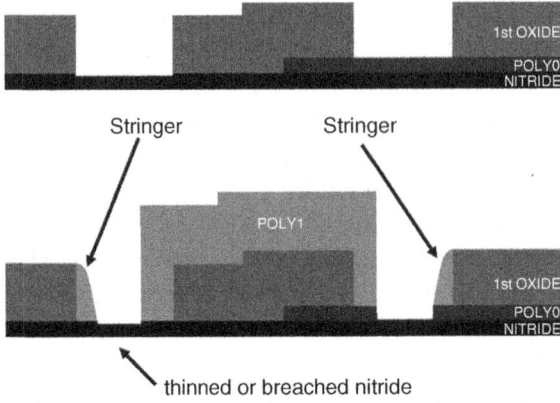

Figure 1.18 Consequences for not following a design rule. (Top) Oxide1 layer patterned with the ANCHOR1 mask. (Bottom) Poly1 patterned with the POLY1 mask. (Reprinted with permission from MEMSCAP Inc.)

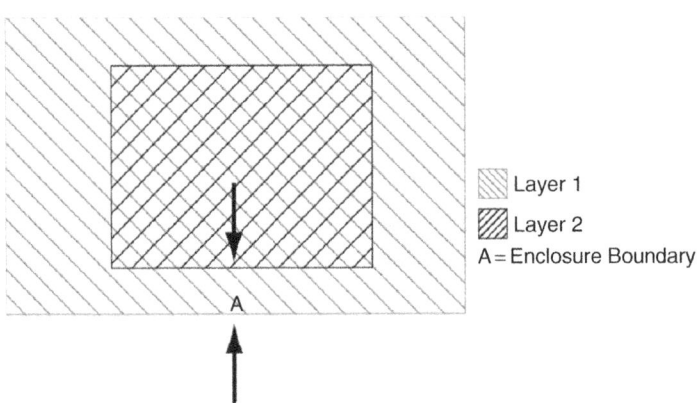

Figure 1.19 Layer 1 surrounds layer 2 by at least a distance A. (Reprinted with permission from MEMSCAP Inc.)

patterned, again using a reactive ion etch with good selectivity between polysilicon and oxide but not good selectivity with nitride. A design rule specifies that features drawn on the mask level POLY1 used to pattern Poly1 should enclose features drawn on the mask ANCHOR1 by at least 4 µm, as shown in Figure 1.19.

In Figure 1.18 this design rule was not followed, giving rise to a number of problems. First, the nitride was thinned because it was exposed during the Poly1 etch and the RIE does not have good selectivity between polysilicon and nitride. Second, the Poly0 layer was completely etched, because it was exposed by the previous ANCHOR1 etch and there is no selectivity between Poly0 and Poly1 during the Poly1 etch. If this layer of Poly0 is meant to be a wire to electrically connect two features, it would be cut during the Poly1 etch and the intended electrical connection will then be open. An additional problem is that the holes in Oxide1 give rise to steps where residual Poly1 remains, giving rise to "stringer" defects that can be unintentionally released during the sacrificial etch step. These "stringers" can form conducting wires that electrically connect features that were meant to be electrically isolated.

In some cases there are minimum design rules and nominal design rules for feature sizes. The minimum design rules push the limits of what can be done, whereas the nominal design rules are well within the limits of what can be done. It is best to use the nominal design rules whenever possible and only to use the minimum design rules when they are absolutely required. Examples include routing wiring through tight spaces and obtaining the minimum spacing for electrostatic actuators when the electrostatic force that is generated depends on the gap spacing. There is a tendency by novices to use the minimum design rules when they are not required. Also, in some cases it may be desirable to violate a design rule to obtain a desired outcome. So long as the consequences do not lead to the failure of other layouts that are included in the multiproject wafer, such as unanchored parts that are released during the sacrificial etch and land on your neighbor's space, you can ignore the design rules. In Chapter 8 we will see a deformable mirror design in which the minimum spacing design rule is intentionally violated to minimize the topography on a conformal overlayer. So long as the consequences can be anticipated, the design rules can be considered as advisory rules rather than mandatory rules.

1.4 Layout

Layout is the activity that is used to define the features on each mask level, and the relationship between features on different mask levels. There are a number of different computer-aided design (CAD) tools available for layout, with some being more user friendly than others. Some examples of layout editors include L-Edit, MEMS Pro, IntelliSense (IntelliMask), AutoCAD, Jale3D, and Coventor (Designer). Some layout editors come with process setup files for the available MPW processes, which can make layouts for these processes much simpler. Some layout tools also include the ability to do automated design rule checking, and to generate solid models and cross-sectional images. These features can be very valuable for generating layouts that lead to working parts on the first try. It is also helpful to sketch solid models and cross sections without using these tools to gain familiarity with a given process.

A typical layout editor is fully hierarchical, with the capability to handle any number of layers, cells with multiple layers, and cells within cells. One word of caution is that when you import a cell from one design into another, to be sure that the cell is copied over rather than externally referenced. A cell that is externally referenced will not be available to another user who is working with your layout. Usually an option is given when importing to either copy or reference the cell. Be sure to copy it rather than referencing it.

The layout editor may have an internal data format that is used to store the layout, and it should also be able to generate output data files in the formats that are commonly used by foundries. These include CIF (Caltech Intermediate Format) and GDSII (Graphic Data System). Some tools do not support these formats, but it may be possible to find converters from one file format, such as DXF (Drawing Exchange Format), to the required format, such as GDSII. Typically the user will transfer the data file to the foundries mask data folder using an FTP (file transfer protocol) command. In some cases the foundry may require the user to submit checksum data to ensure that the file transferred correctly.

Some important features for MEMS layout editors, in comparison to IC layout editors used in microelectronics chip design, are the ability to lay out curved features (e.g., non-Manhattan geometries), because many MEMS devices are not formed from rectangular shapes as are typically used in microelectronic layouts. Another valuable feature is the availability of foundry design kits that include information about the layers and processing steps that are used for various MPW processes. Some examples are

- MEMSCAP's PolyMUMPS, SOIMUMPS, and MetalMUMPS processes [20]
- QinetiQ's INTEGRAMplus Metal Nitride Surface Micromachining (MPK), Deep Etch Silicon-On-Oxide Process (DPK), and 2-Layer Polysilicon (PPK) [21]
- Tronics Microsystems' 60 μm HARM (High Aspect Ratio Micromachining) SOI Process [22]
- Multi-Project Wafer (MPW) for Prototyping on DALSA Semiconductor High Voltage CMOS/DMOS Technologies with MEMS Post-Processing option [23]

1.4 Layout

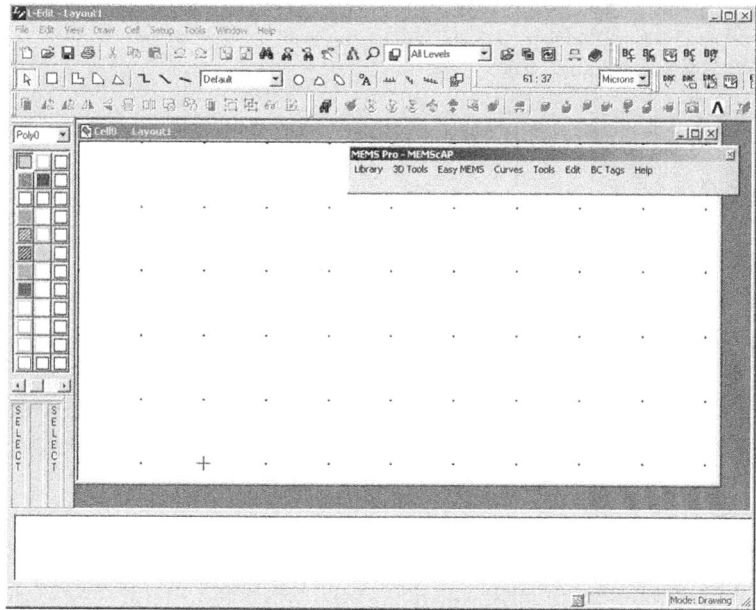

Figure 1.20 L-Edit screen with MEMS Pro upgrade. (Reprinted with permission from Dr. Mary Ann Maher, SoftMEMS.)

- HBSRI – Wafer Dissolved Process [24]
- MicroFabrica's EFAB Multiple Stack Electroplating [25]
- Sandia SUMMiT and iMEMS Processes [26]

Other desirable features include the ability to do three-dimensional (3-d) modeling; view cross sections through the 3-d model; and design rule checking (DRC) and libraries of standard MEMS components such as comb-drives, bond pads, and mechanical test structures.

A screenshot of an L-Edit screen is shown in Figure 1.20. It includes a menu (top), tool bars (top), a layout area (middle), a layer palette (left), and a command line (bottom). In addition, it has location information within the layout (top right).

Before using the layout editor, the technology and grid information should be specified. In L-Edit these are set using the Set-Up Design menu. In the example shown in Figure 1.21, the MUMPS V. 4.1 technology is used, with the technology units set to microns. The L-Edit internal unit specifies the precision of the internal calculations, and the conversion factor specifies the relationship between the internal units and microns. In the setup shown, 1 internal unit = $1/1000\,\mu$m.

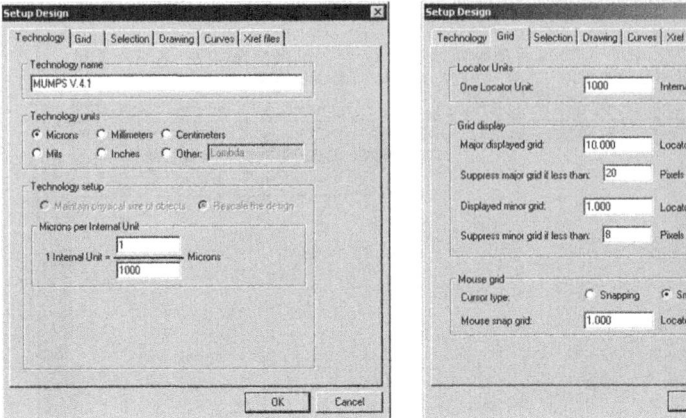

Figure 1.21 Setup design. (Reprinted with permission from Dr. Mary Ann Maher, SoftMEMS.)

Under the "Grid" tab, locator units are specified. Here a locator unit is set equal to 1000 internal units, which has been set equal to 1 µm. The locator bar (top right) in Figure 1.20 displays the location of the mouse pointer and can be used to measure distances by resetting the locator to (0,0) with a right mouse click and measuring relative to this new origin. The grid display can also be set on this tab. Here the size of the major grid has been set to 10 locator units and the size of the minor grid set equal to one locator unit.

The individual layer can also be set up using the Setup Layers option, as shown in Figure 1.22 (right). For the PolyMUMPS process, the layers include Poly0, Anchor1, Dimple, Poly1, Poly1-Poly2 Via, Anchor2, Poly2, and Metal. In addition, holes through each of the poly layers can be defined using Hole0, Hole1, Hole2, and through the metal layer using HoleMetal. These holes aid in defining etch release holes in each of the structural layers. Layers can be added, deleted, copied, renamed, or rearranged. The layer properties can be modified (thickness, stress, resistivity), and the new layers can be derived from existing layers using Boolean operations to merge, substract, grow, and shrink. The colors that are used for rendering the different layers can also be changed, although it is a good idea to keep the renderings set to the default when sharing your layout with others who are familiar with the default colors commonly used in the process.

To the left in Figure 1.22 is the color palette for the different layers. This is similar to the palette of colors used in a painting, and different layers can be selected by a mouse click to change layers. Not all of the layers shown in the color palette or the layer setup are necessarily used.

1.4 Layout

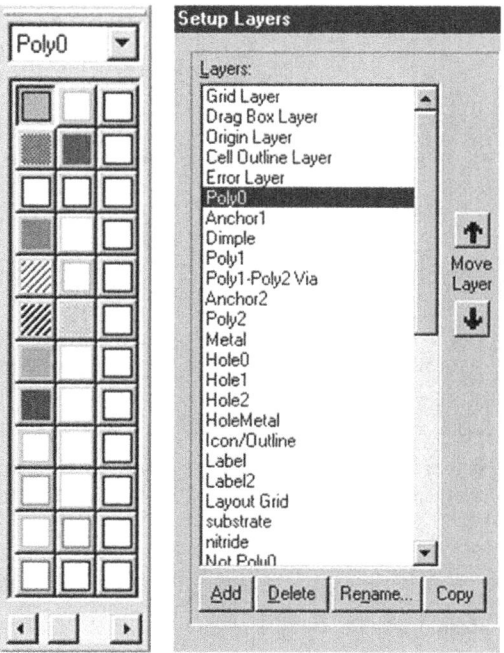

Figure 1.22 Color palette and layer setup. (Reprinted with permission from Dr. Mary Ann Maher, SoftMEMS.) See color plate section.

Figure 1.23 Drawing toolbar. (Reprinted with permission from Dr. Mary Ann Maher, SoftMEMS.)

The drawing toolbar, shown in Figure 1.23, can be used to select different shapes that will be drawn in the layout using the layer that has been selected from the color palette. Drawing objects include boxes, polygons, circles, pie wedges, toruses, ports, rulers, and instances. A right mouse click on the drawing toolbar can be used to toggle between different drawing options such as "orthogonal," "all angle," and "curves."

Some of the shapes that can be drawn are shown in Figure 1.24. The shapes include polygons (90 degrees, 45 degrees, all-angle), wires (90 degrees, 45 degrees, all-angle), standard curves (circles, tori, pie wedges), and special curves (splines, ellipses, spirals, sinusoids, hyperbolas, parabolas).

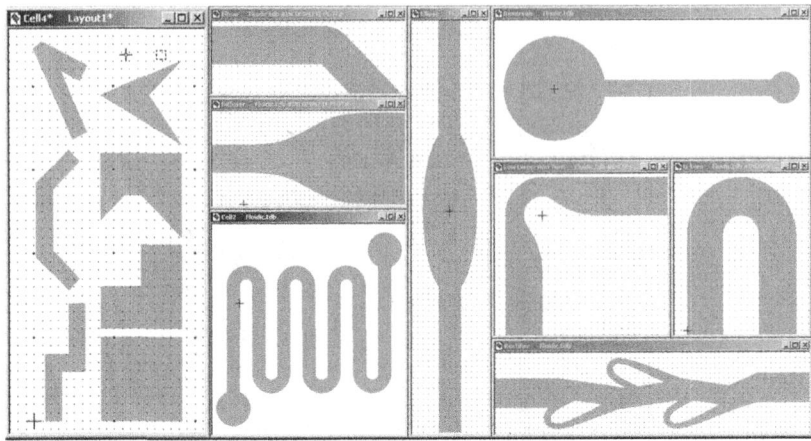

Figure 1.24 Shapes that can be drawn using the drawing toolbar. (Reprinted with permission from Dr. Mary Ann Maher, SoftMEMS.)

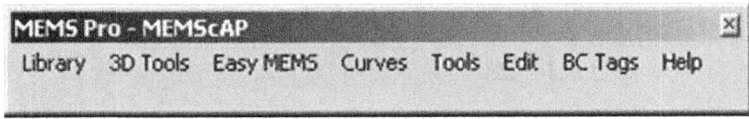

Figure 1.25 MEMS Pro toolbar. (Reprinted with permission from Dr. Mary Ann Maher, SoftMEMS.)

The MEMS Pro toolbar is shown in Figure 1.25. The MEMS Pro software is an upgrade to the L-Edit software that can make layouts much easier. It includes library elements for standard MEMS components such as comb-drives, motors, springs, bearings, test structures, resonator elements, and fluidic elements. It also includes standard designs for bond pads. The use of standard cells for design is highly encouraged to avoid mistakes. A common mistake is to leave out an anchor on a bond pad so that it gets released during the sacrificial etch, as shown in Figure 1.9. Not only do you lose a bond pad for making a connection to your MEMS design, your neighbors on the multiproject wafer may gain a bond pad they did not necessarily desire!

An example of the library elements for active elements, passive elements, and resonator elements is shown in Figure 1.26. This library includes shuttle plates, folded springs, rotary torsional springs, linear comb-drives and rotary comb-drives, rotary motors, and test circuits. Having these standard components available at the click of a mouse can significantly accelerate layout.

1.4 Layout

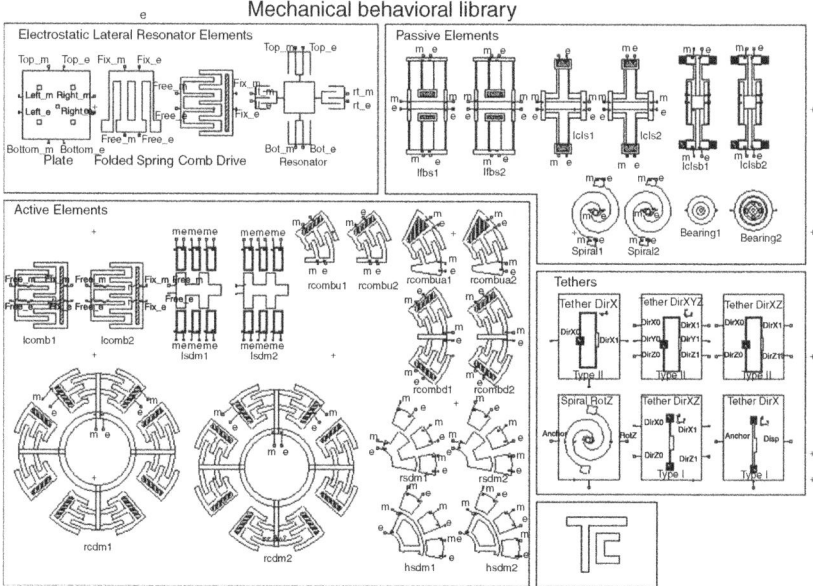

Figure 1.26 MEMS standard cells. The use of standard cells can increase productivity in layout while avoiding layout errors by using known good designs. (Reprinted with permission from Dr. Mary Ann Maher, SoftMEMS.)

Problems

(1) Use L-Edit/MEMS Pro to lay out your name in the PolyMUMPS process. Use the Poly0 layer to make the letters, which should be about 100 μm tall and 50 μm wide. Once you have made your name, make a 100 × 100 array of your name. Copy and paste the layouts into your homework. What layers could you use if you wanted to release the letters from the substrate to make an alphabet soup?

(2) Use L-Edit/MEMS Pro to lay out each of the M-Test structures (cantilever beam, fixed-fixed beam, and clamped circular diaphragm) in the PolyMUMPS process as shown schematically in Figure 1.27 [27]. Use Poly0 for the ground plane and Poly1 for the released structure in each layout. Use Oxide1 as the sacrificial layer to release each of the structural elements. The figure shows a dielectric spacer to electrically isolate the released structure from the ground plane so that the structure can be actuated by applying a voltage V. What layer in the PolyMUMPS process can be used to electrically isolate a

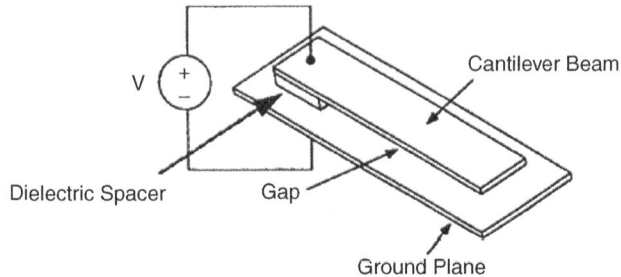

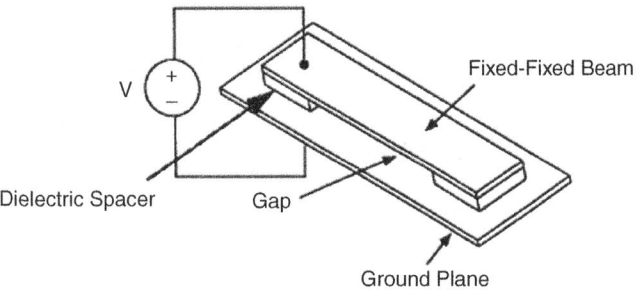

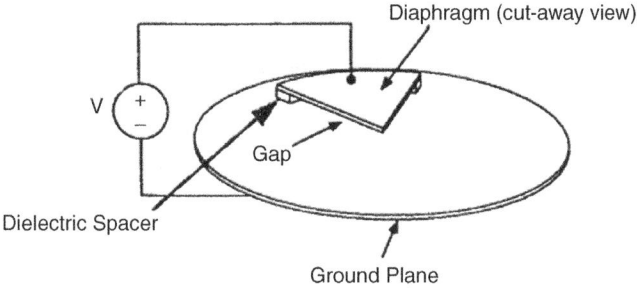

Figure 1.27 M-Test structures (cantilever beam, fixed-fixed beam, diaphragm). (Reprinted with permission from Journal of Microelectromechanical Systems, *M-TEST: a test chip for MEMS material property measurement using electrostatically actuated test structures*, P.M. Osterberg and S.D. Senturia, ©1967 IEEE.)

released structure in Poly1 from a ground plane defined in Poly0? If you use oxide, be sure that it is not completely etched away during the sacrificial release step or your structure will be released! Use the cross-section tool in MEMS Pro to find a cross section of the structure including the anchor to the substrate. Make a solid model

1.4 Layout

of each structure. Experiment with the scaling in the z-direction. Copy and paste your results (layouts, cross sections, and solid models) into your homework.

(3) Lay out a released wheel in the PolyMUMPS process using Poly1 for the wheel and Poly 2 for the hub that constrains the wheel and keeps it from falling off of the chip [4], [5]. You should not need to use any Poly1 for the hub! Use the cross-section tool to find a cross section of the wheel that goes through the hub. Copy and paste your results (layout and cross section) into your homework.

(4) Lay out a hinge in the PolyMUMPS process using Poly1 for the hinge pin and Poly2 for the hinge staple [6], [7] as shown in Figure 1.28. Use the cross-section tool to find a cross section of the hinge that goes through the hinge pin along line A–A' in Figure 1.28. Copy and paste your results (layout and cross section) into your homework.

(5) Identify 10 different methods to produce a mechanical force on a microscale. Determine the scaling laws for each of your methods.

(6) Read the article "There's plenty of room at the bottom" by Richard Feynman [28]. As you read through the article, make a list of the predictions he made for Microsystems. What did he get right? What did he get wrong?

(7) Read the article "Infinitesimal machinery" by Richard Feynman [29]. As you read through the article, make a list of the predictions he made for Microsystems. What did he get right? What did he get wrong?

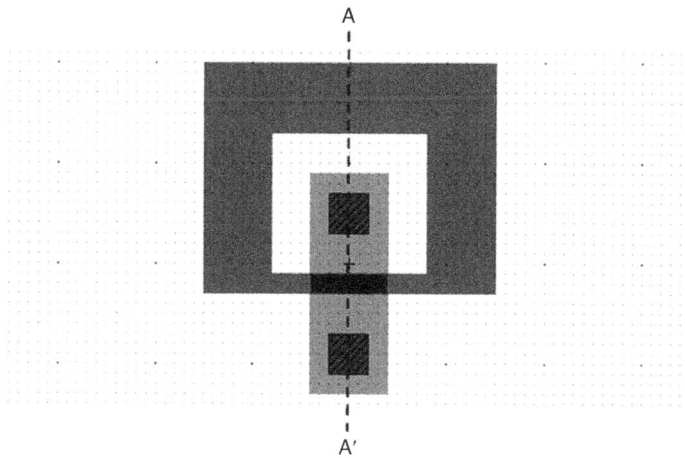

Figure 1.28 Hinge.

REFERENCES

1. http://www.tms.org/pubs/journals/JOM/0509/Chen-0509.html (last accessed on 3–17–2010).
2. H.C. Nathanson, W.E. Newell, R.A. Wickstrom, and J.R. Davis Jr., *The resonant gate transistor*, IEEE Trans. Electron Devices 14, p. 117 (1967).
3. K.W. Markus and K.J. Gabriel, *MEMS: The systems function revolution*, IEEE Trans. Electron Devices 14, pp. 25–31 (1999).
4. R.T. Howe and R.S. Muller, *Polycrystalline silicon micro-mechanical beams*, Proc. Electrochemical Society Spring Meeting, Montreal, Quebec, Canada, May 9–14, pp. 184–185 (1982).
5. J.M. Bustillo, R.T. Howe, and R.S. Muller, *Surface micromachining for microelectromechanical systems*, Proc. IEEE 86(8), pp. 1552–1574 (1998).
6. http://www.memscap.com/en_mumps.html (last accessed on 3–17–2010).
7. G.K. Fedder, *MEMS fabrication*, ITC International Test Conference, pp. 691–698 (2003).
8. J. Carter, A. Cowen, B. Hardy, R. Mahadevan, M. Stonfield, and S. Wilcenski, *PolyMUMPS Design Handbook*, Revision 11.0 (2005).
9. L.-S. Fan; Y.-C. Tai, and R.S. Muller, *IC-processed electrostatic micro-motors*, International Electron Devices Meeting, IEDM '88. Technical Digest, pp. 666–669 (1988).
10. Y.-C. Tai, L.-S. Fan, and R.S. Muller, *IC-processed micro-motors: Design, technology, and testing*, Micro Electro Mechanical Systems, 1989, Proceedings, An Investigation of Micro Structures, Sensors, Actuators, Machines and Robots, IEEE 20–22, pp. 1–6 (1989).
11. K.S.J. Pister, M.W. Judy, S.R. Burgett, and R.S. Fearing, *Microfabricated hinges: 1 mm vertical features with surface micromachining*, IEEE Transducers '91, San Francisco, CA, pp. 647–650 (1991).
12. K.S.J. Pister, M.W. Judy, S.R. Burgett, and R.S. Fearing, *Microfabricated hinges*, Sensor and Actuators A 33(3), pp. 249–256 (1992).
13. V.A. Aksyuk, F. Pardo, and D.J. Bishop, *Stress-induced curvature engineering in surface micromachined devices*, SPIE 3680, pp. 984–993 (1999).
14. G.T. Mulhern, D.S. Soane, and R.T. Howe, *Supercritical carbon dioxide drying of microstructures*, Proc. Transducers '93 (Yokohama), pp. 296–299 (1993).
15. P.J. French and P.M. Sarro, *Surface versus bulk micromachining: The contest for suitable applications*, J. Micromechanical Microengineering 8, pp. 45–53 (1998).
16. J.J. Sniegowski, S.M. Rodgers, B.G. Boone, J.R. Bruzzi, C.W. Drabenstadt, B.E. Kluga, E.W. Rogala, R. Osiander, K.J. Rebello, and M.A.G. Darrin, *Development, test and evaluation of MEMS micro-mirrors for free-space optical communications*, Proc. SPIE 5550, pp. 299–312 (2004).
17. N. Doble, M. Helmbrecht, M. Hart, and T. Juneau, *Advanced wavefront correction technology for the next generation of adaptive optics equipped ophthalmic instrumentation*, Proc. SPIE 5688, pp. 125–132 (2005).
18. R. Wood, V. Dhuler, and E. Hill, *A MEMS variable optical attenuator*, Proc. IEEE/LEOS Int'l Conf. Opt. MEMS, Kauai, HI, Aug. 2000, pp. 121–222 (2000).

19 J.A. Kubby on behalf of the MOEMS Manufacturing Consortium, *Hybrid integration of light emitters and detectors with SOI-based micro-opto-electro-mechanical (MOEMS) systems*, Proc. SPIE 4293, Silicon-Based and Hybrid Optoelectronics III, D.J. Robbins, J.A. Trezza, G.E. Jabbour, Eds., pp. 32–45 (2001).
20 http://www.memscap.com/en_mumps.html (last accessed on 3–17–2010).
21 http://www.qinetiq.com/home/capabilities/electronics/INTEGRAMplus/Smart_Silicon_MEMS_Prototyping_and_Manufacture.html (last accessed on 3–17–2010).
22 http://it-assistant.eu/mbcd/tronic_1.html (last accessed on 3–17–2010).
23 http://www.dalsa.com/public/semi/DES-0041.05.pdf (last accessed on 3–17–2010).
24 http://www.mems-issys.com/dissolved.shtml (last accessed on 3–17–2010).
25 http://www.memgen.com/pages/index.php (last accessed on 3–17–2010).
26 http://www.mems.sandia.gov/tech-info/summit-v.html (last accessed on 3–17–2010).
27 P.M. Osterberg and S.D. Senturia, *M-TEST: A test chip for MEMS material property measurement using electrostatically actuated test structures*, J. Microelectromechanical Systems 6(2), pp. 107–118 (1997).
28 R.P. Feynman, *There's plenty of room at the bottom*, reprinted in J. Microelectromechanical Systems 1(1), pp. 60–66 (1992).
29 R.P. Feynman, *Infinitesimal machinery*, reprinted in J. Microelectromechanical Systems 2(1), pp. 4–14 (1993).

2
Micromechanics

2.1 Springs

Microelectromechanical systems (MEMS) respond to forces that act on them. The forces can either be external to the MEMS device, such as the measurement of acceleration forces in an accelerometer, or from internal forces, such as an electrostatic force in an actuator. In many cases the resulting deformations can be calculated using one-dimensional linear spring analysis based on Hooke's Law:

$$F = -k\delta. \tag{2.1}$$

Here the amount of deflection of the spring δ is proportional to the force F with the proportionality constant being the spring constant k. The spring constant k is the ratio of the stimulus F to the response δ. More generally, the amount by which a material body is deformed, the strain ε, is linearly related to the force causing the deformation, the stress σ. If we consider a beam of length L and cross-sectional area A subjected to a tensile force F, the beam will be lengthened by an amount ΔL according to the relations

$$\begin{aligned} \sigma &= \frac{F}{A} \quad \frac{N}{m^2} \text{ (Pa)}, \\ \varepsilon &= \frac{\Delta L}{L}, \\ \sigma &= E\varepsilon. \end{aligned} \tag{2.2}$$

Here E, the constant of proportionality between the stress σ (stimulus) and strain ε (response), is called the modulus of elasticity, or Young's modulus. E is a property of the material that indicates how "stiff" it is. Young's modulus for polysilicon is approximately 160 GPa. Young's modulus for single crystal silicon depends on the crystallography. E is

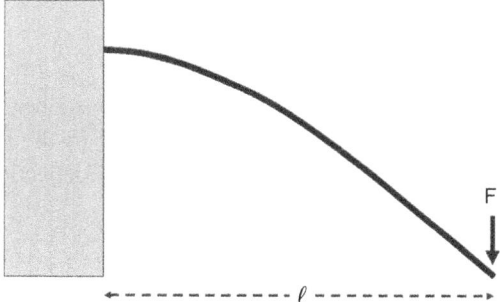

Figure 2.1 A cantilever beam of length l subjected to a point force F applied perpendicular to its length.

169 GPa parallel to the wafer flat for a (100) oriented wafer and 130 GPa at 45 degrees to the wafer flat. The relation between stress and strain can be put into the same form as Hooke's Law, relating the lengthening (shortening) of the beam to the tensile (compressive) force acting on it:

$$\sigma = E\varepsilon,$$
$$\frac{F}{A} = E\frac{\Delta L}{L},$$
$$F = \left(\frac{EA}{L}\right)\Delta L, \qquad (2.3)$$

where the effective spring constant $k = EA/L$. Let us consider how much a polysilicon beam is stretched with different forces acting on it. If the beam is 100 μm long, 5 μm wide, and 2 μm thick and is subjected to a tensile force of 1 μN,

$$\Delta L = \left(\frac{L}{EA}\right)F = \left(\frac{100 \times 10^{-6}\,\text{m}}{(160 \times 10^9\,\text{Pa})(2 \times 10^{-6}\,\text{m})(5 \times 10^{-6}\,\text{m})}\right)(1 \times 10^{-6}\,\text{N})$$
$$= 6.25 \times 10^{-11}\,\text{m}$$
$$= 0.0625\,\text{nm}. \qquad (2.4)$$

Clearly the beam is very stiff in tension and does not stretch significantly, so it does not make a very effective spring. Nonetheless, a beam can be used as a spring if the force is applied perpendicular to the length of the beam, as shown in Figure 2.1.

A beam of length L that is clamped at one end and subjected to a bending force F acting on the free end will deflect according to

$$\delta = \left(\frac{L^3}{3EI}\right)F \quad \text{where } I = \frac{wt^3}{12}. \quad (2.5)$$

Here I is the moment of inertia of the beam, where the beam has a width $w > t$ and the force F is pushing along the normal to the thin direction. For our example polysilicon beam, we would get a deflection δ:

$$\delta = \left(\frac{(100 \times 10^{-6} \text{ m})^3}{3(160 \times 10^9 \text{ Pa})}\left(\frac{12}{(5 \times 10^{-6} \text{ m})(2 \times 10^{-6} \text{ m})^3}\right)\right)(1 \times 10^{-6} \text{ N})$$
$$= 6.25 \times 10^{-7} \text{ m}$$
$$= 0.625 \text{ μm}. \quad (2.6)$$

which is clearly a much more effective spring! The spring constant k is given by

$$k = \frac{F}{\delta} = \frac{3EI}{L^3} = \left(\frac{3E}{L^3}\right)\left(\frac{wt^3}{12}\right)$$
$$= \left(\frac{3(160 \times 10^9 \text{ Pa})}{(100 \times 10^{-6} \text{ m})^3}\right)\left(\frac{(5 \times 10^{-6} \text{ m})(2 \times 10^{-6} \text{ m})^3}{12}\right)$$
$$= \frac{1 \times 10^{-6} \text{ N}}{0.625 \times 10^{-6} \text{ m}} = 1.6 \text{ N/m}. \quad (2.7)$$

2.1.1 Springs connected in parallel

For springs that are connected in parallel, as shown in Figure 2.2, the deflection of each spring, Δx, would be the same [1]:

$$F_{total} = F_1 + F_2$$
$$= -k_1 \Delta x_1 - k_2 \Delta x_2$$
$$= -(k_1 + k_2)\Delta x \quad (2.8)$$
$$= -k_{eff_{parallel}} \Delta x$$
$$k_{eff_{parallel}} = k_1 + k_2.$$

2.1.2 Springs connected in series

For springs that are connected in series, as shown in Figure 2.3, the force acting on each of the springs would be the same; but because they have

2.2 Buckling

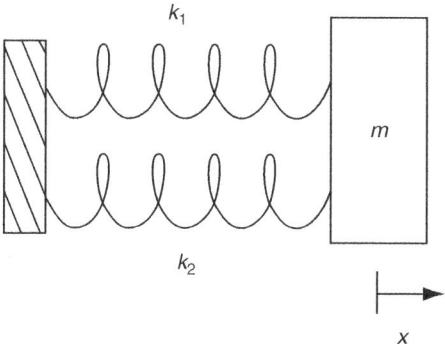

Figure 2.2 Two springs connected in parallel. One spring has a spring constant k_1 and the other spring has a spring constant k_2. Both are stretched by the same amount Δx [1].

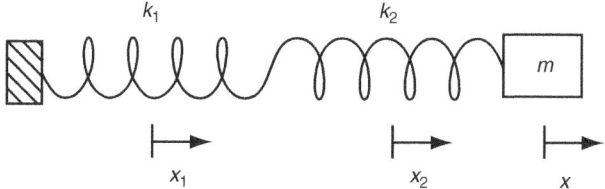

Figure 2.3 Two springs connected in series. One spring has a spring constant k_1 and is stretched by a distance Δx_1, and the other has a spring constant k_2 and is stretched by a distance Δx_2. The same force F acts on both springs [2].

different spring constants (stiffnesses), they would stretch by different amounts, Δx_1 and Δx_2 [2]:

$$\Delta x = \Delta x_1 + \Delta x_2 = \frac{F_1}{k_1} + \frac{F_2}{k_2} = \left(\frac{1}{k_1} + \frac{1}{k_2}\right)F$$
$$\frac{\Delta x}{F} = \frac{1}{k_{\textit{eff series}}} = \frac{1}{k_1} + \frac{1}{k_2}.$$
(2.9)

2.2 Buckling

What happens if we put a compressive force on a beam, such as using it to push something? If it reaches a critical level of compressive stress, called the Euler buckling limit, the beam will buckle. The critical stress is given by:

$$\sigma_{Euler} = -\frac{\pi^2}{3}E\left(\frac{t}{L}\right)^2. \tag{2.10}$$

It is worthwhile to see what compressive force would be required for our polysilicon beam to buckle:

$$\frac{F_{critical}}{A} = \sigma_{Euler} = -\frac{\pi^2}{3}E\left(\frac{t}{L}\right)^2$$

$$F_{critical} = -\frac{\pi^2}{3}EA\left(\frac{t}{L}\right)^2 = -\frac{\pi^2}{3}(160 \times 10^9 \text{ Pa})(5 \times 10^{-6} \text{ m})$$

$$(2 \times 10^{-6} \text{ m})\left(\frac{2 \times 10^{-6} \text{ m}}{100 \times 10^{-6} \text{ m}}\right)^2$$

$$= 2 \text{ mN}. \tag{2.11}$$

As we shall see, forces of this magnitude can be easily generated with thermal actuators, and the critical force decreases with the square of the beam length. *In general, it is better to use beams to pull rather than to push on objects to avoid buckling the beam!*

2.3 Poisson's ratio

In general, if we put a compressive stress on an object, it will decrease in length. Alternatively, if we put a tensile stress on an object, it will elongate. However, stresses imposed in one direction can lead to changes in dimensions in other directions. The tendency can be thought of as the preservation of volume, as shown in Figure 2.4. A familiar example of this is the finger trap shown in Figure 2.5. As you elongate the tube, its diameter shrinks, trapping your fingers! Poisson's ratio v is a measure of this tendency. Poisson's ratio is the ratio of the relative contraction strain, or the transverse strain (normal to the applied load), divided by the relative extension strain, or the axial strain (in the direction of the applied load):

$$v = -\frac{\varepsilon_{transvere}}{\varepsilon_{axial}} = -\frac{\varepsilon_x}{\varepsilon_y}, \tag{2.12}$$

where ε_x is the transverse strain (negative for axial tension, positive for axial compression) and ε_y is the axial strain (positive for axial tension, negative for axial compression) [3].

2.3 Poisson's ratio

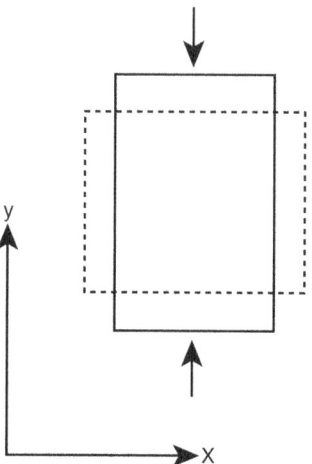

Figure 2.4 Rectangular specimen subject to compression, with Poisson's ratio $\upsilon \approx 0.5$. (Reprinted with permission from Janet Kozicki at the English Wikipedia project.)

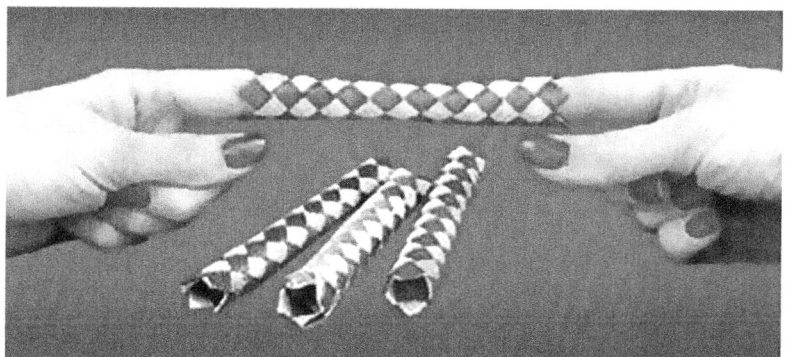

Figure 2.5 As you pull your fingers apart you elongate the tube and shrink the diameter, trapping your fingers tighter the harder you pull! (Reprinted with permission from the Wikimedia Commons.)

Most materials have υ between 0.0 and 0.5. Cork is close to 0.0, polysilicon is around 0.22, single crystal silicon is around 0.28, most steels are around 0.3, and rubber is almost 0.5. A perfectly incompressible material deformed elastically at small strains would have a Poisson's ratio of exactly 0.5. Some materials, mostly polymer foams, have a negative Poisson's ratio; if these "auxetic" materials are stretched in one direction, they become thicker in perpendicular directions.

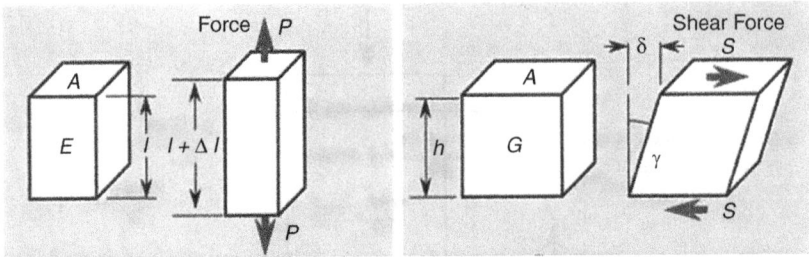

Figure 2.6 Comparison of axial stress and strain (left) with shear stress and strain (right). (Reprinted with permission from Prof. Hiroshi Toshiyoshi, Institute of Industrial Science (IIS), The University of Tokyo, Japan.)

2.4 Shear stress and strain

In addition to deflecting a beam by pushing with a normal force P on one of its faces, it is also possible to cause a beam to deflect by applying a shearing force S parallel to the surface of an object, as shown in Figure 2.6. The anchor applies an equal and opposite shearing force. The shear stress τ is given by

$$\tau = \frac{S}{A} \quad \left(\frac{\mathrm{N}}{\mathrm{m}^2}\right). \tag{2.13}$$

The shearing force S causes the beam to tilt at an angle γ, called the shear strain, where

$$\gamma = \frac{\delta}{h} \quad \text{(radians)}. \tag{2.14}$$

The stimulus–response relationship between the shear stress and shear strain is given by the shear modulus G:

$$\tau = G\gamma. \tag{2.15}$$

It can be seen that there is a correspondence between the axial stress and strain and the shear stress and strain:

$$\begin{array}{cc} \text{Axial} & \text{Shear} \\ \sigma = \dfrac{F}{A} & \tau = \dfrac{S}{A} \\ \varepsilon = \dfrac{\Delta l}{l} & \gamma = \dfrac{\delta}{h} \\ \sigma = E\varepsilon & \tau = G\gamma \end{array} \tag{2.16}$$

For an isotropic medium such as polysilicon, the shear modulus G and Young's modulus E are related by

$$G = \frac{E}{2(1+v)}, \tag{2.17}$$

where v is Poisson's ratio.

2.5 Beams in other situations

Beams are often subject to different boundary conditions and forces in MEMS design. A comprehensive reference for the bending of beams in different situations is Roark's formulas for stress and strain [4]. We considered a fixed-free cantilever beam that is subjected to a point load at its free end in the spring analysis above. The fixed-free boundary condition means that one end is fixed, so that both its displacement and slope do not change under the applied force. The other end of the beam is free to both move and change its slope in response to a point load F, as shown in Figure 2.7. In that case the deflection of the beam $y(x)$ as a function of position x is given by

$$y(x) = \frac{Fx^2}{6EI}(3l - x) \quad I = \frac{wt^3}{12}. \tag{2.18}$$

In some cases the end of a cantilever beam is not free to both displace and change its slope, but rather the slope is fixed by an attached structure such as a released membrane, as shown in Figure 2.8. The nonfixed end of

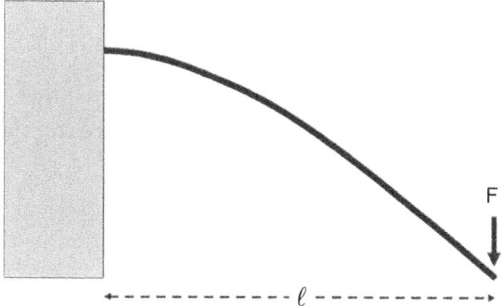

Figure 2.7 Bending of a fixed-free cantilever beam subjected to a point load F at the free end.

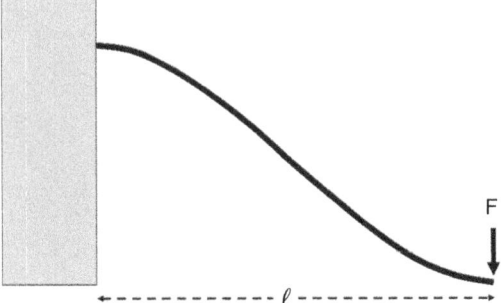

Figure 2.8 Fixed-guided cantilever beam subjected to a point load F at the guided end.

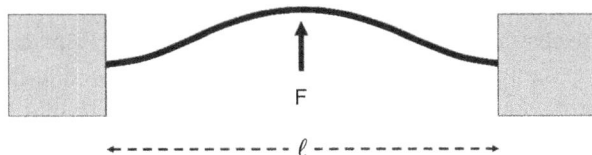

Figure 2.9 Fixed-fixed beam subjected to a point load F at the middle of the beam.

the beam is still able to be displaced, but the slope on the free end is fixed by its attachment to something that is stiffer than the beam. This sort of boundary condition is called a fixed-guided boundary condition. In this case the deflection y(x) is given by

$$y(x) = \frac{Fx^2}{12EI}(3l - 2x). \tag{2.19}$$

If both ends of the beam are fixed (*fixed-fixed*) as shown in Figure 2.9 and a point load F is applied to the middle of the beam, the deflection is given by

$$y(x) = \frac{Fx^2}{48EI}(3l - 4x) \quad 0 \leq x \leq \frac{l}{2} \tag{2.20}$$

If a distributed load $\rho = F/L$ is applied along the length of a fixed-free beam, as shown in Figure 2.10, the deflection is given by

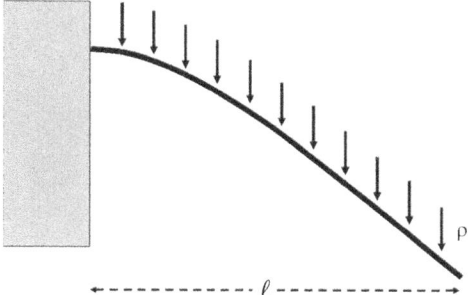

Figure 2.10 Fixed-free cantilever beam subjected to a distributed load ρ along the length of the beam.

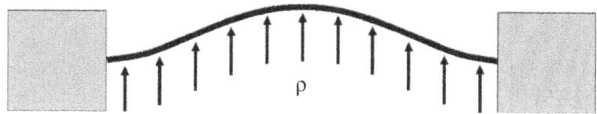

Figure 2.11 Fixed-fixed beam subjected to a distributed load ρ along the length of the beam.

$$y(x) = \frac{\rho x^2}{24EI}\left(6l^2 - 4lx + x^2\right). \tag{2.21}$$

If a distributed load ρ is applied along the length of a fixed-fixed beam, as shown in Figure 2.11, the deflection is given by

$$y(x) = \frac{\rho x^2}{24EI}(l - x)^2. \tag{2.22}$$

2.6 Torsion

In addition to linear springs, which apply a linear restoring force when they are stretched, a number of MEMS designs also make use of torsional springs, which apply a restoring torque when twisted. A familiar example is the torsion rods that support the tip-tilt mirror in the Texas Instruments (TI) Digital Light Processing (DLP) project display. A schematic diagram of a torsion rod is shown in Figure 2.12.

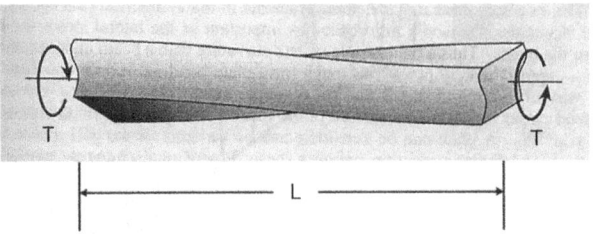

Figure 2.12 Schematic diagram of a torsion rod of length L that is twisted by an applied torque T (from Gregory T.A. Kovacs, *Micromachined Transducers Sourcebook*, McGraw-Hill, Boston, 1998, ISBN 0-070290722-3).

The twist angle θ in radians can be found from

$$\theta = \frac{TL}{KG}, \tag{2.23}$$

where T is the applied torque, L is the length of the torsion rod, K is the torsional moment of inertia, and G is the shear modulus of elasticity. A table of torsional moments can be found in Rourke's formulas for stress and strain. For a torsion rod with a circular cross section with radius r, the torsional moment is given by

$$K = \frac{1}{2}\pi r^4. \tag{2.24}$$

For a torsion rod with a square cross section where the sides are of length $2a$, the torsional moment is given by

$$K = 2.25 a^4. \tag{2.25}$$

For a rectangular cross section with sides of length $2a$ and $2b$, where $a > b$, the torsional moment is given by

$$K = ab^3 \left[\frac{16}{3} - 3.36 \frac{b}{a}\left(1 - \frac{b^4}{12a^4}\right) \right] \quad \text{for } a \geq b. \tag{2.26}$$

2.7 Membranes

In addition to beams, MEMS devices such as pressure sensors also use membranes that are clamped around the edges. The deflections

2.8 Test structures

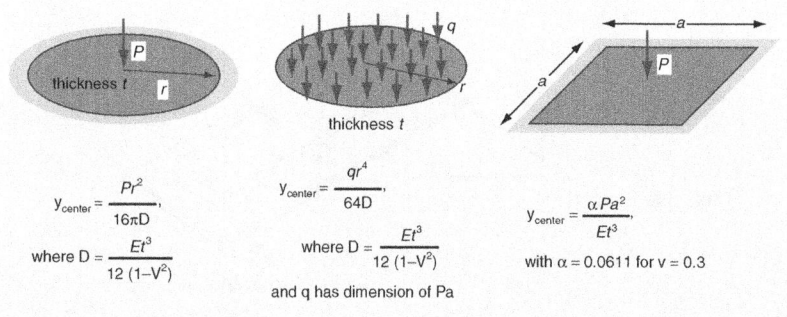

Figure 2.13 Deflection of released membranes. (Left) Clamped circular diaphragm with thickness t and radius r subjected to a point load P at its center. (Middle) Clamped circular diaphragm subjected to a distributed load q N/m^2 or Pa. (Right) Clamped square diaphragm with sides of length a subjected to a point load P at its center. (Reprinted with permission from Prof. Hiroshi Toshiyoshi, Institute of Industrial Science (IIS), The University of Tokyo, Japan.)

of membranes subjected to both point forces and distributed loads (pressure) are shown in Figure 2.13.

2.8 Test structures

It is important to include test structures in the layout to measure materials' properties such as Young's modulus, residual stress, and stress-gradients that can vary with the processing conditions used to fabricate the MEMS devices. We will focus on a set of simple mechanical test structures called M-Test structures [5]. Analogous to the electrical test structures (E-Test) that are included in microelectronic wafers to measure process-dependent materials properties like polysilicon resistivity and field effect transistor thresold voltages, the M-Test structures will allow mechanical properties like stiffness, stress, and stress-gradients to be determined. The M-Test structures are shown in Figure 2.14. They consist of cantilever beams (CB), fixed-fixed beams (FB), and clamped circular diaphragms (CD) of various sizes.

As we shall see in Chapter 3 on electrostatics, these structures can be electrostatically actuated and display a sharp threshold phenomenon called a "pull-in" instability, where the released structure is pulled down to the ground plane when a certain voltage threshold has been achieved. The pull-in can be observed with a microscope, so that expensive test

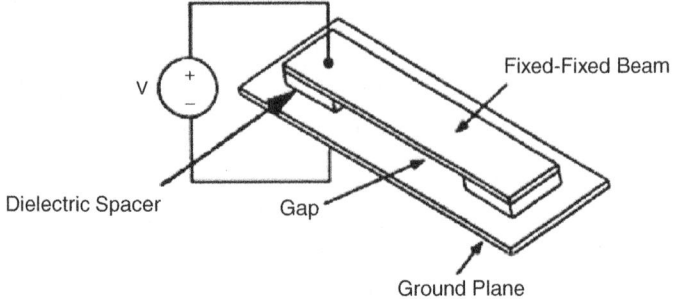

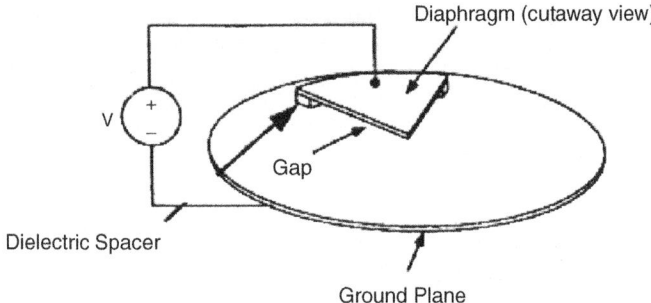

Figure 2.14 M-Test structures. (Top) Fixed-fixed beam (FB). (Bottom) Clamped diaphragm (CD). (Reprinted with permission from J. Microelectromechanical Systems, *M-TEST: a test chip for MEMS material property measurement using electrostatically actuated test structures*, P.M. Osterberg and S.D. Senturia, ©1967 IEEE.)

equipment is not required, just a variable power supply and a voltmeter. A closed-form solution can be obtained for the pull-in voltage for each of these structures that allows a stress parameter S and a bending parameter B to be determined. These parameters, along with the geometric dimensions of the test structures, enable residual stress and the Young's modulus to be determined. In addition, the cantilever beams will indicate stress-gradients through the thickness of the film because a stress-gradient that is tensile on top will cause the cantilever to bend up away from the substrate and a stress-gradient that is compressive on top will cause the cantilever to bend down toward the substrate, as shown in Figure 1.5. An array of fixed-fixed beams (FB) can also indicate residual compressive stress because they will buckle once the Euler criteria has been met. An example of an array of cantilever beams and fixed-fixed beams is shown in Figure 2.15 [6].

2.8 Test structures

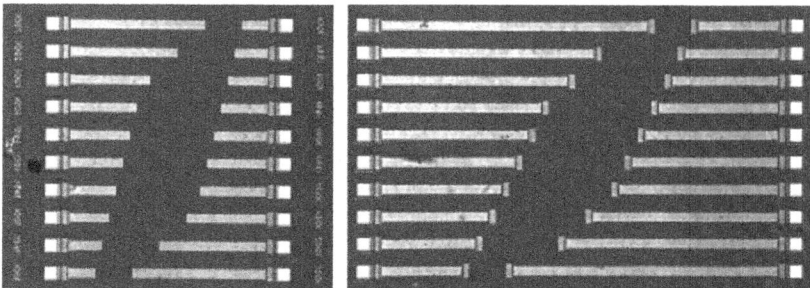

Figure 2.15 Photographs showing arrayed cantilever (left) and fixed-fixed (right) beam test structures fabricated by polysilicon surface micromachining. Fixed-fixed beam lengths range from 300 µm to 1000 µm, and cantilevers from 100 µm to 500 µm. (Reprinted with permission from Raj K. Gupta, Ph.D. dissertation, Massachusetts Inst. Technol., Cambridge, MA, 1997.)

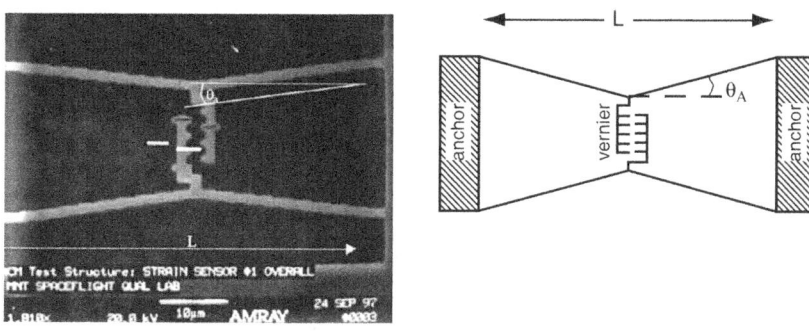

Figure 2.16 Bent-beam strain sensor for characterization of residual stress. Under tensile stress the arms will deflect outward. Under compressive stress the arms will deflect inward. (Reprinted with permission from J. Microelectromechanical Systems, *Bent-beam strain sensors*, Y. B. Gianchandani and K. Najafi, ©1967 IEEE.)

Another useful test structure is the bent-beam strain sensor that is used to measure residual tensile stress. An example is shown in Figure 2.16 [7], [8]. Similar to a guitar string, the bent beam will tend to straighten under tensile residual stress, pulling the vernier outward. The bent beam will tend to bend further under a compressive residual stress, pushing the vernier inward. By including bent-beam strain sensors oriented in orthogonal directions, the directional dependence of the residual stress can be measured.

2.9 Dampening

When microelectromechanical systems move in air, the viscous loss (friction) can provide dampening for the device. A familiar example of dampening in macromechanical systems is the oil dashpot that is included in a mass–spring system, as shown in Figure 2.17 (left), which is described by the second-order differential equation

$$m\frac{d^2x}{dt^2} + c\frac{dx}{dt} + kx = 0. \tag{2.27}$$

The dampening coefficient c is the viscous dampening coefficient for a velocity dependent force:

$$F = -cv = -c\frac{dx}{dt}. \tag{2.28}$$

The motion of a MEMS device through air can give rise to a viscous force called squeezed film dampening. As shown in Figure 2.17 (right), the MEMS device can squeeze the air film trapped in a narrow gap

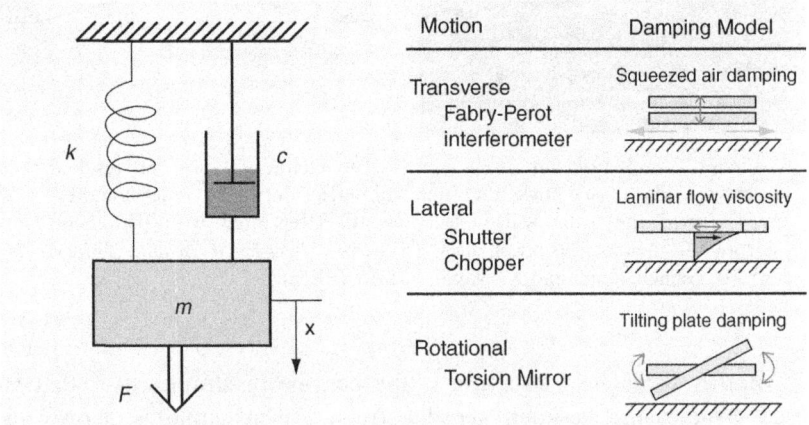

Figure 2.17 (Left) Mass (m)–spring (k) system with dampening (c). (Right) Different forms of dampening depending on the motion of the released MEMS component. (Top) Transverse motion squeezes the air. (Middle) Lateral motion gives rise to a viscous force. (Bottom) Rotational motion causes both squeezed film dampening and viscous dampening. (Reprinted with permission from Prof. Hiroshi Toshiyoshi, Institute of Industrial Science (IIS), The University of Tokyo, Japan.)

between a released member and the substrate, like a piston; move through air, like a wing; or force air from one position to another, like a compressor. Each of these different situations is shown in Figure 2.17 (right). If the air is compressed rather than displaced, the air film can act as a spring.

2.10 Accelerometer

An accelerometer is essentially a mass on a spring. The mass is released from the substrate and is supported by springs. The released mass m reacts to an external force F according to

$$F_{inertia} = ma = m\frac{d^2x}{dt^2}. \qquad (2.29)$$

If the mass is supported by springs with spring constant k, the springs supply an elastic restoring force:

$$F_{elasticity} = -kx. \qquad (2.30)$$

And if the mass moves through a viscous medium like air at a velocity v, there will be a viscous force:

$$F_{viscous} = -cv = -c\frac{dx}{dt}, \qquad (2.31)$$

where c is the dampening coefficient for air. For our initial analysis we will consider a static acceleration so that there is no viscous force. The force balance equation is then given by

$$m\frac{d^2x}{dt^2} = kx. \qquad (2.32)$$

2.10.1 Cantilever beam

We first consider the deflection of a cantilever beam subjected to an acceleration a. The deflection δ at the end of a cantilever beam of length L under a distributed load F/L (N/m) is

$$\delta = \frac{\rho L^4}{8EI}. \qquad (2.33)$$

If the cantilever has a mass m that is subjected to an acceleration ng, the load would then be nmg and the deflection would be

$$\delta = \left(\frac{nmg}{L}\right)\frac{L^4}{8EI}. \tag{2.34}$$

As an example, we consider a polysilicon cantilever beam that is 1 mm long, 20 μm wide, and 2 μm thick. The mass of the beam would be

$$m = \rho V = \left(2330 \ \frac{\text{kg}}{\text{m}^3}\right)(1 \times 10^{-3} \ \text{m})(20 \times 10^{-6} \ \text{m})(2 \times 10^{-6} \ \text{m})$$

$$= 9.32 \times 10^{-11} \ \text{kg}. \tag{2.35}$$

The deflection, in terms of g, would then be

$$\frac{\delta}{g} = nm\frac{L^3}{8EI} = nm\left(\frac{L^3}{8E}\right)\left(\frac{12}{wt^3}\right)$$

$$= n(9.32 \times 10^{-11} \ \text{kg})\left(\frac{(1 \times 10^{-3} \ \text{m})^3}{8 \times 160 \times 10^9 \ \text{Pa}}\right)\left(\frac{12}{(20 \times 10^{-6} \ \text{m})(2 \times 10^{-6} \ \text{m})^3}\right)$$

$$= n(5.46 \times 10^{-3} \ \mu\text{m}). \tag{2.36}$$

Under an acceleration of $n = 100$ g, the cantilever beam would deflect by 0.55 μm.

2.10.2 Crash sensor

An example of an automotive crash sensor that could be used to deploy an air-bag safety system is shown in Figure 2.18. Here a proof mass is suspended by four fixed-guided support arms of length L that are attached to a stiff frame. The frame is rigidly attached to the car.

The acceleration a of the proof mass is given by

$$a = \frac{k}{m}x. \tag{2.37}$$

For fixed-guided cantilever beams, the spring constant for each beam would be

$$k = \frac{12EI}{L^3} = \frac{12E}{L^3}\frac{wt^3}{12} = Ew\left(\frac{t}{L}\right)^3$$

$$k_{total} = 4k = 4Ew\left(\frac{t}{L}\right)^3. \tag{2.38}$$

2.10 Accelerometer

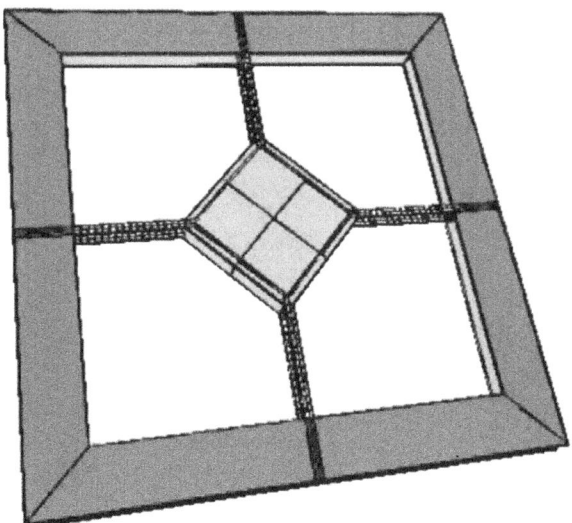

Figure 2.18 Crash sensor. (Reprinted with permission of Sandeep Akkaraju from IntelliSuite v8.6 (2010), IntelliSense Software Corporation.)

If we have the support arms made from Poly2 in the PolyMUMPS process that are 1.5 μm thick, 10 μm wide, and 1 mm long, the spring constant would be

$$k_{total} = 4k = 4Ew\left(\frac{t}{L}\right)^3$$
$$= 4(160 \times 10^9)(10 \times 10^{-6} \text{ m})\left(\frac{1.5 \times 10^{-6} \text{ m}}{1 \times 10^{-3} \text{ m}}\right)^3$$
$$= 0.0216 \, \frac{\text{N}}{\text{m}}. \tag{2.39}$$

If the center plate is formed from a 1 mm² stack of Poly1 (2 μm) and Poly2 (1.5 μm), the mass would be

$$m = \rho V = \left(2330 \, \frac{\text{kg}}{\text{m}^3}\right)(1 \times 10^{-3} \text{ m})^2 (3.5 \times 10^{-6} \text{ m})$$
$$= 8.2 \times 10^{-9} \text{ kg}. \tag{2.40}$$

The deflection of the plate δ would then be

$$\delta = \left(\frac{m}{k}\right)a = \left(\frac{8.2 \times 10^{-9} \text{ kg}}{0.0216 \, \frac{\text{N}}{\text{m}}}\right)(ng) \tag{2.41}$$
$$\frac{\delta}{g} = 0.38 \, \frac{\mu\text{m}}{g}.$$

2.11 Pressure sensor

We next consider a small (linear) deflection of a clamped circular diaphragm under a distributed load as an example of a pressure sensor. The deflection δ at the center of a stress-free clamped circular diaphragm with radius r is given by

$$\delta = \frac{Pr^4}{64D} \quad D = \frac{Et^3}{12(1-v^2)}, \qquad (2.42)$$

where D is the flexural rigidity of the membrane. The deflection of a circular polysilicon membrane that is 1.5 μm thick (e.g., Poly2 of the PolyMUMPS process) with a radius $r = 50$ μm under 1 atmosphere of pressure (101 325 Pa) difference from the front to back would be

$$\delta = \frac{(101\,325 \text{ Pa})(50 \times 10^{-6} \text{ m})^4}{64D}$$

$$D = \frac{(160 \times 10^9 \text{ Pa})(1.5 \times 10^{-6} \text{ m})^3}{12(1-(0.22)^2)}$$

$$= 4.73 \times 10^{-8}$$

$$= 2.3 \text{ μm}. \qquad (2.43)$$

For a square membrane with sides of length L, the deflection δ at the center of the membrane would be

$$\delta = 0.0138 \frac{PL^4}{Et^3}. \qquad (2.44)$$

An example of a clamped square membrane with a piezoresistor around the edge to detect the strain from the membrane deflection is shown in Figure 2.19. The membrane and piezoresistor have been meshed for finite element analysis.

For a 1.5 μm thick polysilicon membrane with sides that are 100 μm long,

$$y = 0.0138 \frac{(101\,325 \text{ Pa})(100 \times 10^{-6} \text{ m})^4}{(160 \times 10^9 \text{ Pa})(1.5 \times 10^{-6} \text{ m})^3} = 0.26 \text{ μm}. \qquad (2.45)$$

The stress at the center of a long edge would be

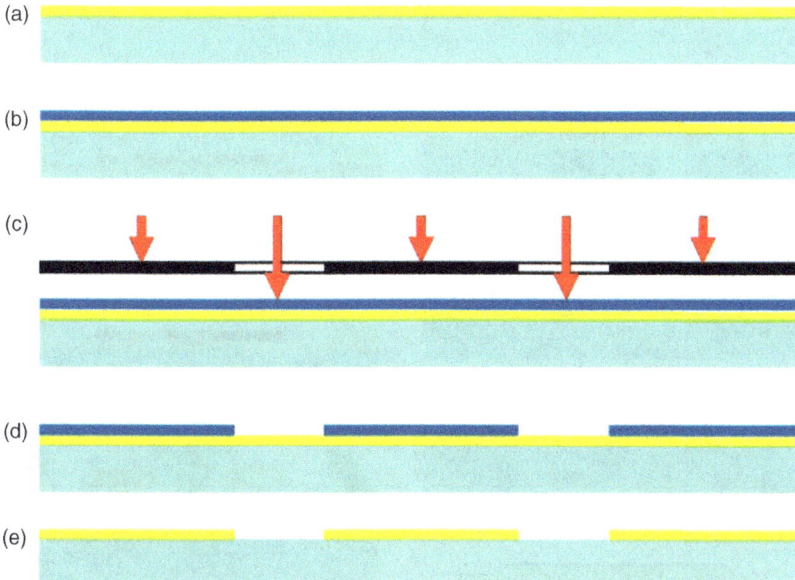

Figure 1.1 The semiconductor fabrication process. (a) Thin-film deposition (yellow), (b) photoresist deposition (blue), (c) photolithography (mask clear and opaque; red arrows), (d) photoresist development, and (e) etching to transfer the pattern in the photoresist into the thin film.

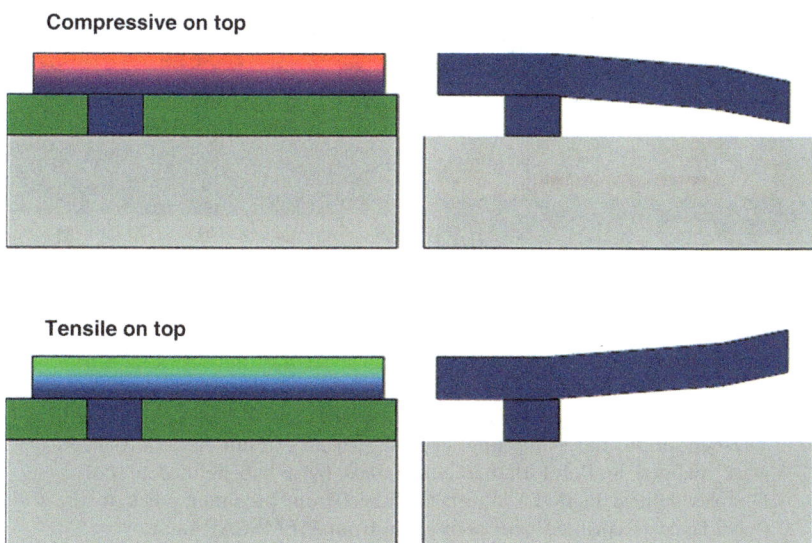

Figure 1.5 Deformation of released structures due to residual stress-gradients. In the top figure the stress is compressive on top, resulting in a downward deflection on release. In the bottom figure the film is tensile on top, resulting in an upward deflection on release.

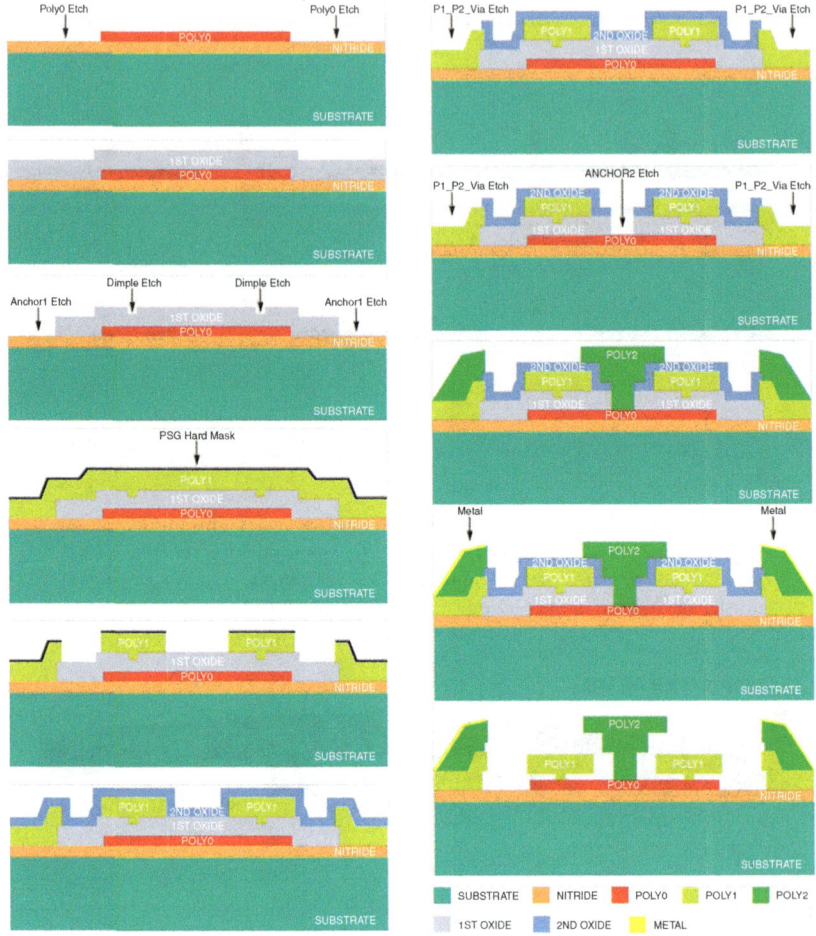

Figure 1.7 PolyMUMPS three-layer polysilicon surface micromachining process offered by MEMSCAP. Polysilicon and oxide layers are deposited and patterned in a cyclic process, with anneal steps of the doped sacrificial oxide between polysilicon depositions. Poly0 is an electrical layer that is not released. Poly1 and Poly2 are structural layers that can be released. The deposition and patterning steps shown here result in a polysilicon wheel defined in Poly1 that is constrained by a hub defined in Poly2. Dimples defined in POLY1 keep the wheel from becoming stuck to the Poly0 layer. (Reprinted with permission from MEMSCAP Inc.)

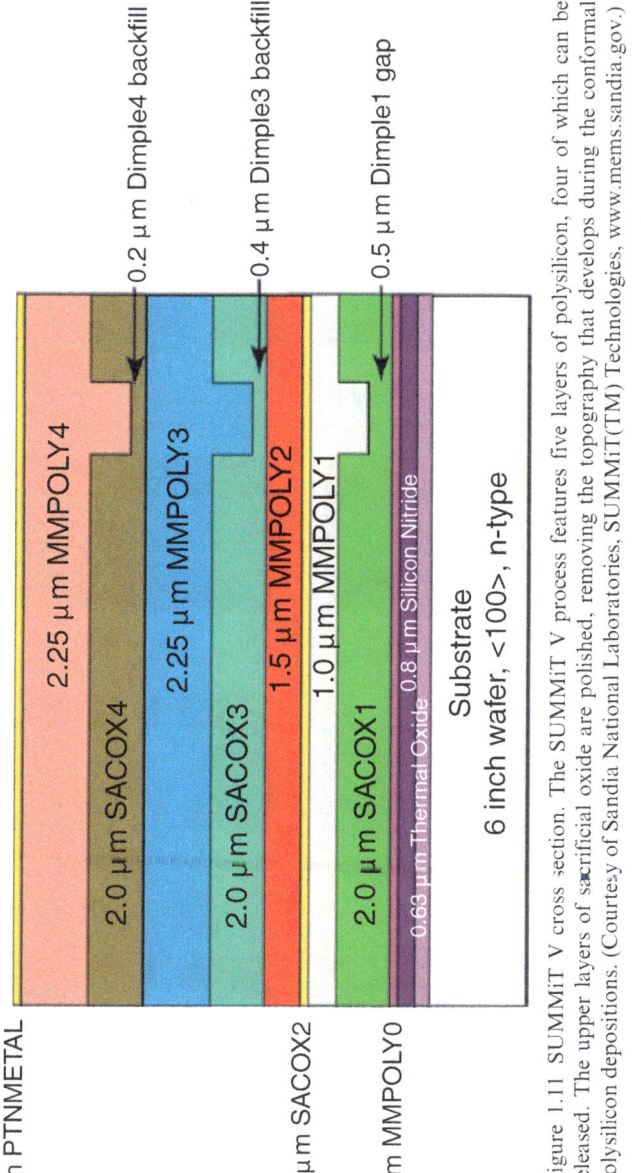

Figure 1.11 SUMMiT V cross section. The SUMMiT V process features five layers of polysilicon, four of which can be released. The upper layers of sacrificial oxide are polished, removing the topography that develops during the conformal polysilicon depositions. (Courtesy of Sandia National Laboratories, SUMMiT(TM) Technologies, www.mems.sandia.gov.)

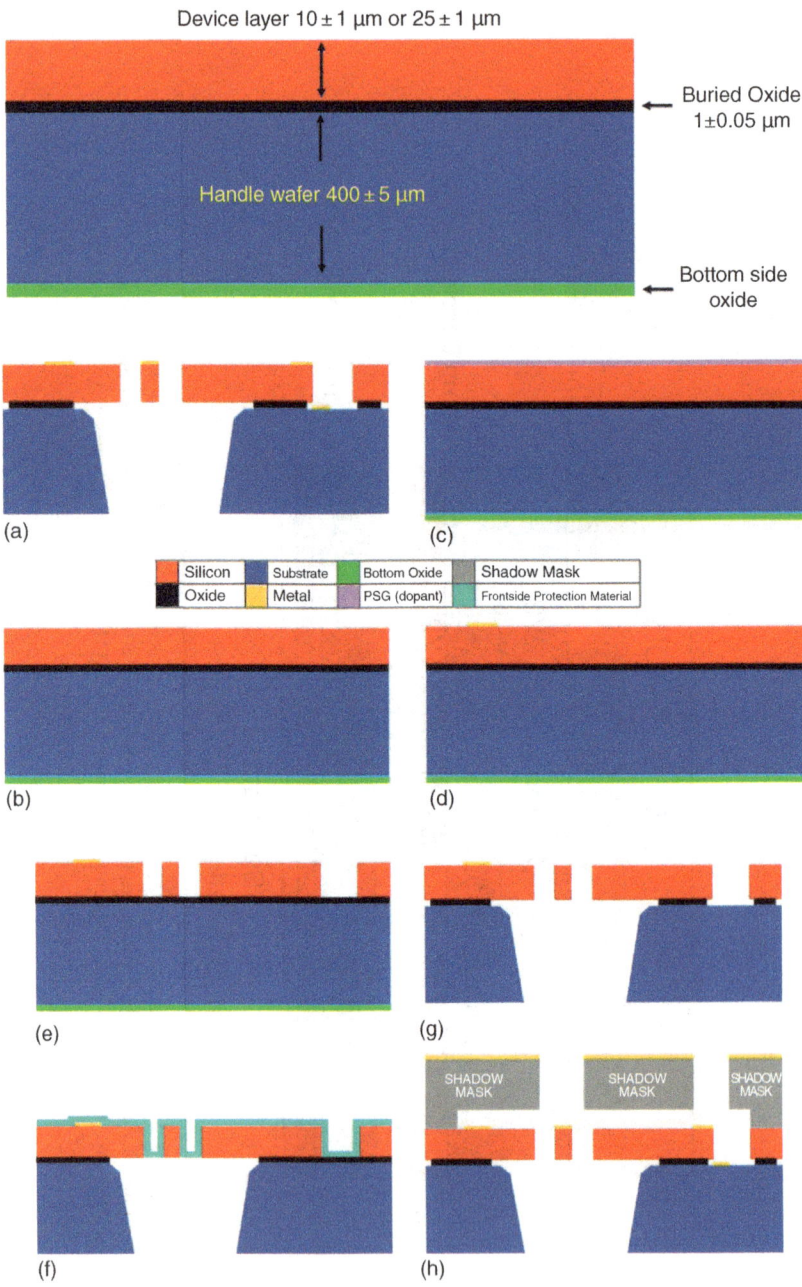

Figure 1.12 Silicon on insulator (SOI) wafer that is used in the SOIMUMPS process. The device layer can be 10 ± 1 μm or 25 ± 1 μm thick. The handle wafer is 400 ± 5 μm thick and the buried oxide layer is 1 ± 0.05 μm thick. (Reprinted with permission from MEMSCAP Inc.)

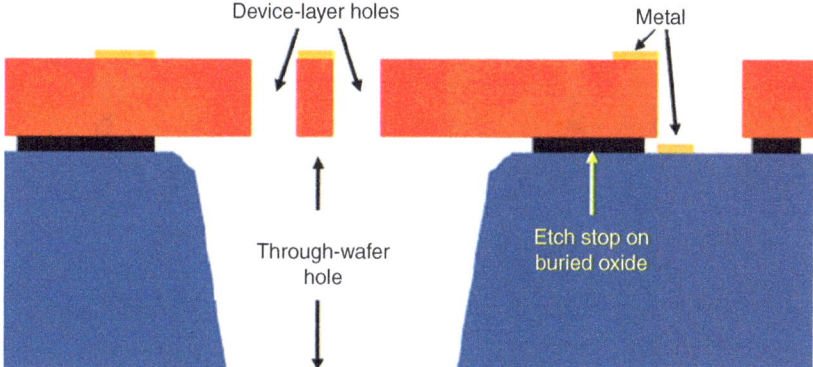

Figure 1.13 Patterned SOIMUMPS wafer. Through-wafer etches are performed from the front side of the wafer 10 or 25 μm deep to form device layer holes, and from the back side 400 μm deep to form through-wafer holes. Both etches stop on the buried oxide. (Reprinted with permission from MEMSCAP Inc.)

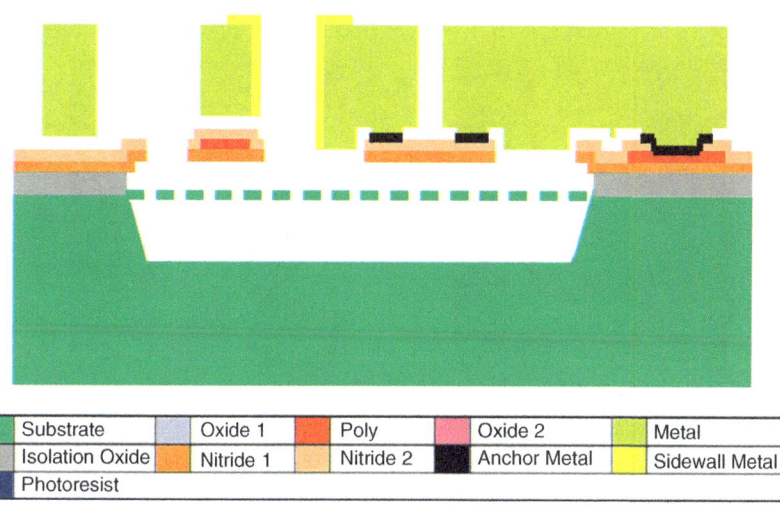

Substrate	Oxide 1	Poly	Oxide 2	Metal
Isolation Oxide	Nitride 1	Nitride 2	Anchor Metal	Sidewall Metal
Photoresist				

Figure 1.16 Cross section through the layer stack used to fabricate the microrelay. (Reprinted with permission from MEMSCAP Inc.)

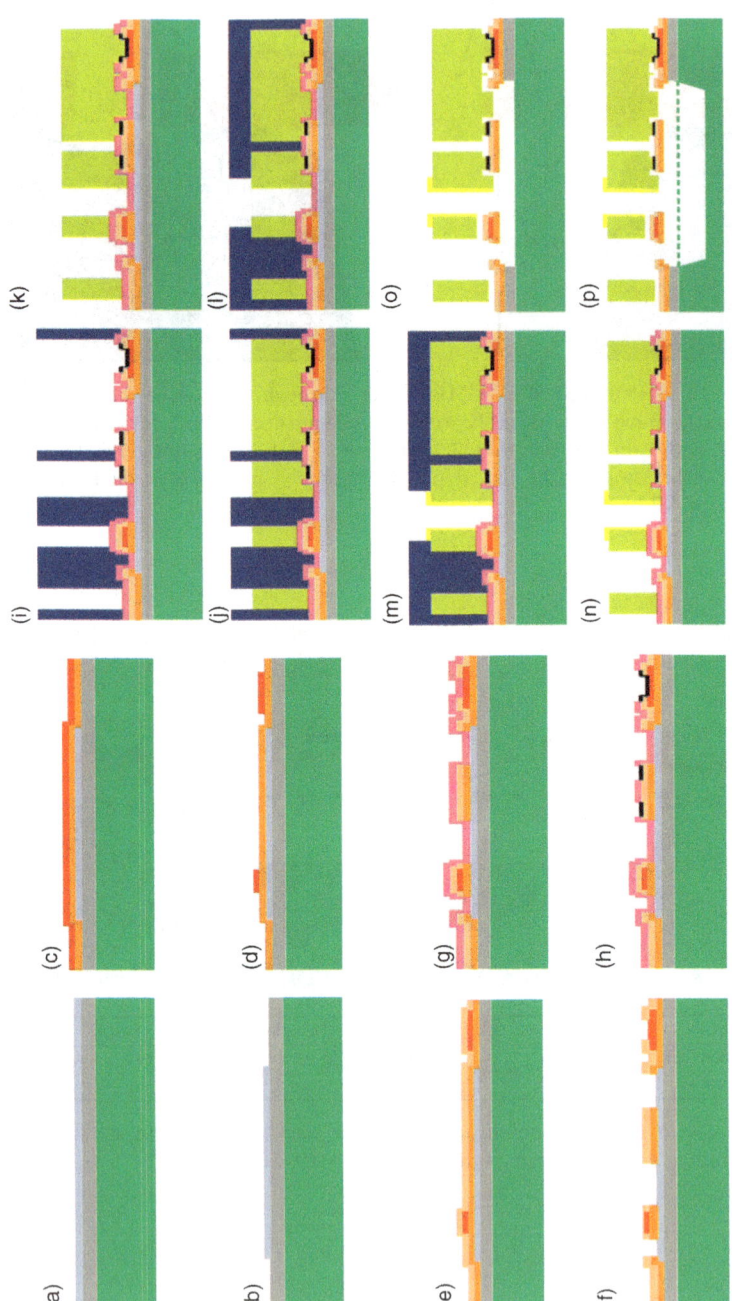

Figure 1.17 Cross sections through the MetalMUMPS process. (Reprinted with permission from MEMSCAP Inc.)

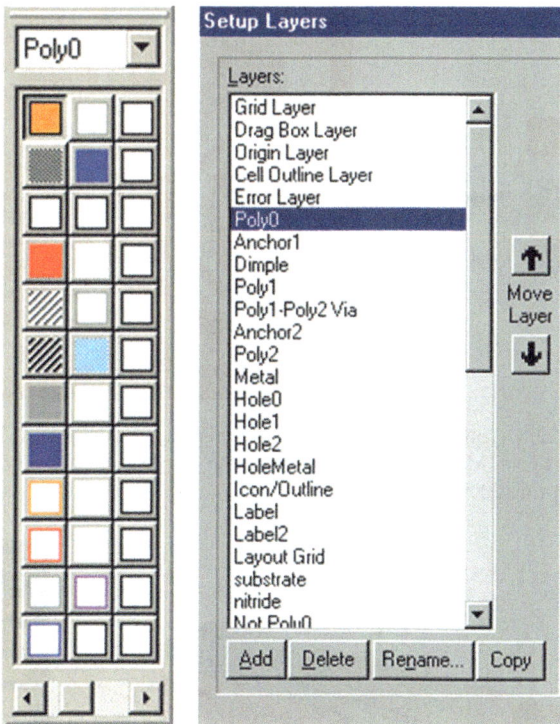

Figure 1.22 Color palette and layer setup. (Reprinted with permission from Dr. Mary Ann Maher, SoftMEMS.)

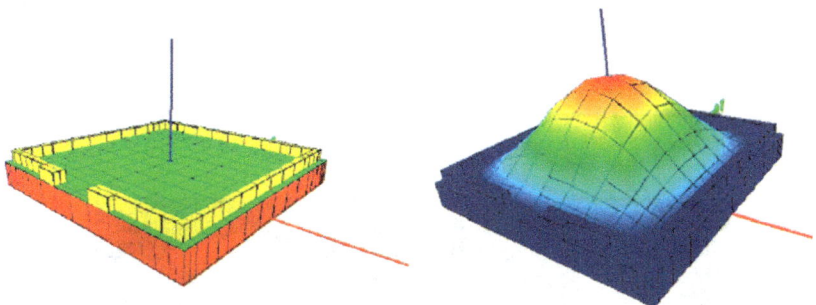

Figure 2.19 Clamped square membrane with a pressure P applied to the back face. (Left) Undeflected membrane. (Right) Deflected membrane with color keyed to displacement. (Reprinted with permission of Sandeep Akkaraju from IntelliSuite v8.6 (2010), IntelliSense Software Corporation.)

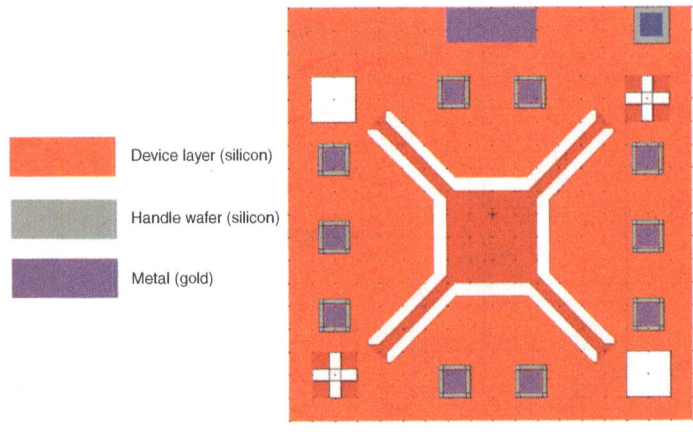

Figure 2.22 Layout for a Fabry-Perot interferometer in the SOIMUMPS process. (From Mr. Dmitry Kozak and the MEMS Design students in EE115, Spring 2009.)

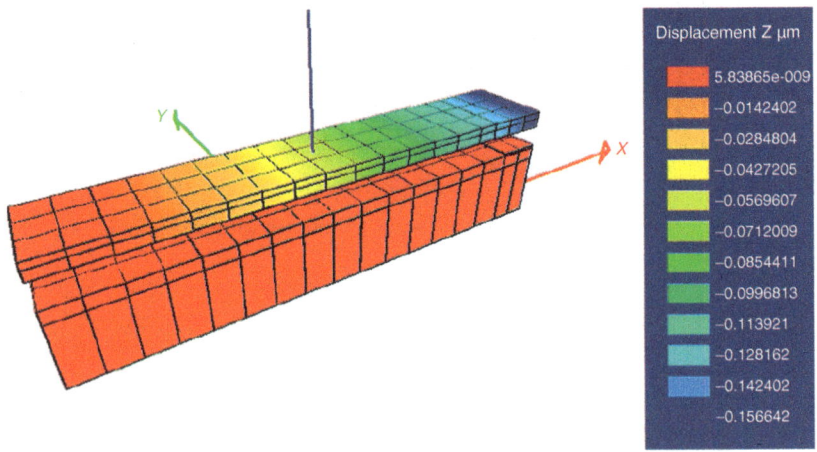

Figure 4.2 Deflection of an electrostatically actuated cantilever beam solved in IntelliSuite. (Reprinted with permission of Sandeep Akkaraju from IntelliSuite v8.6, 2010, IntelliSense Software Corporation.)

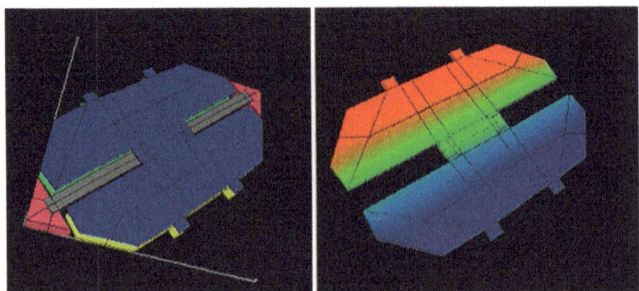

Figure 4.6 Coupled electromechanical modeling of a MEMS torsional mirror. (Reprinted with permission of Sandeep Akkaraju from IntelliSuite v8.6 (2010), IntelliSense Software Corporation.)

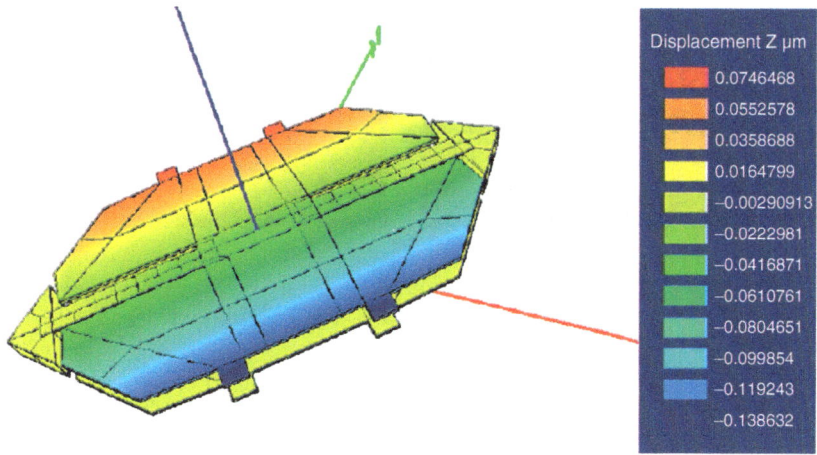

Figure 4.7 Displacement for a MEMS torsional mirror. (Reprinted with permission of Sandeep Akkaraju from IntelliSuite v8.6, 2010, IntelliSense Software Corporation.)

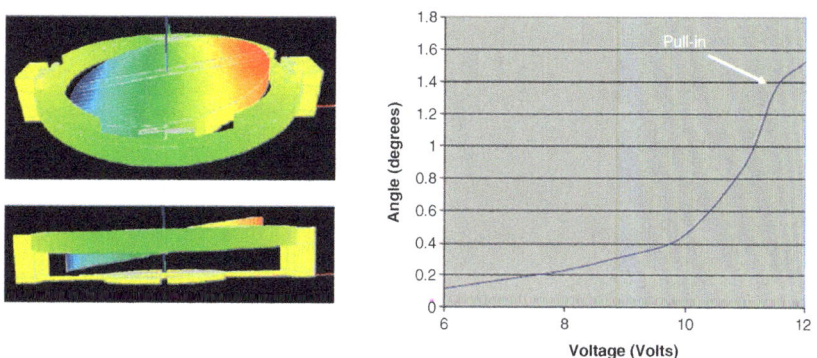

Figure 4.10 Tilt of the mirror about one axis. The deflection of the mirror is shown in the solid model on the left-hand side of the figure and the tilt angle as a function of the applied voltage is shown on the right-hand side of the figure. The pull-in voltage is at approximately 11.5 V, as shown by the arrow in the figure. (Image courtesy of Kuan-Fu Chen and Tung-Chien Chen, EE215 MEMS Design final project report, Spring 2007.)

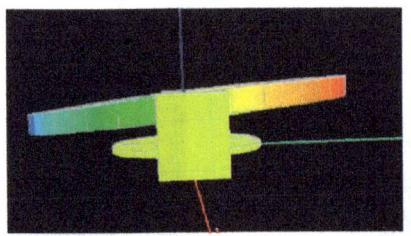
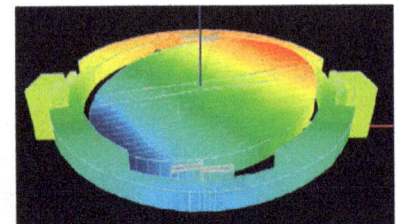

Figure 4.11 (Left) Tip of the mirror by actuation of the ring. (Right) Tip and tilt of the mirror by actuating both the mirror and the ring. (Image courtesy of Kuan-Fu Chen and Tung-Chien Chen, EE215 MEMS Design final project report, Spring 2007.)

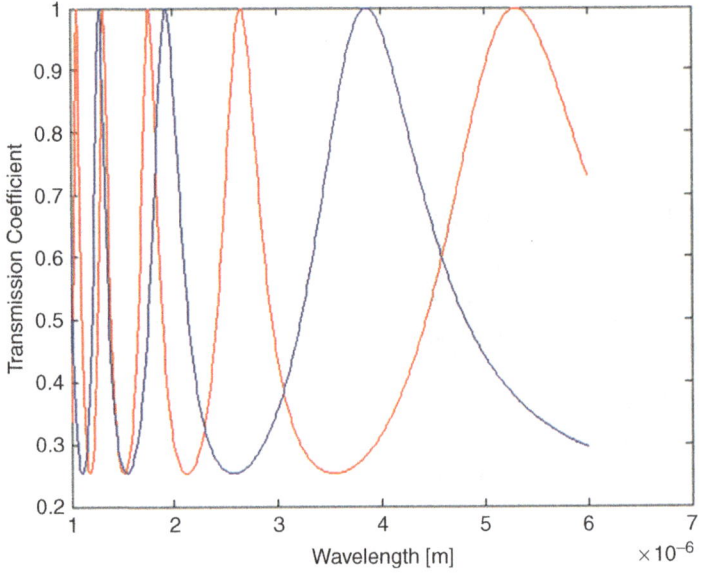

Figure 4.17 The shift in the transmission coefficient of the mid-wavelength IR filter calculated with Matlab as a function of the wavelength. (Courtesy of Kevin Louchis and Benjamin Hemphill, Team Ninja Star in the Eye, EE115-Winter 2008, Final Report.)

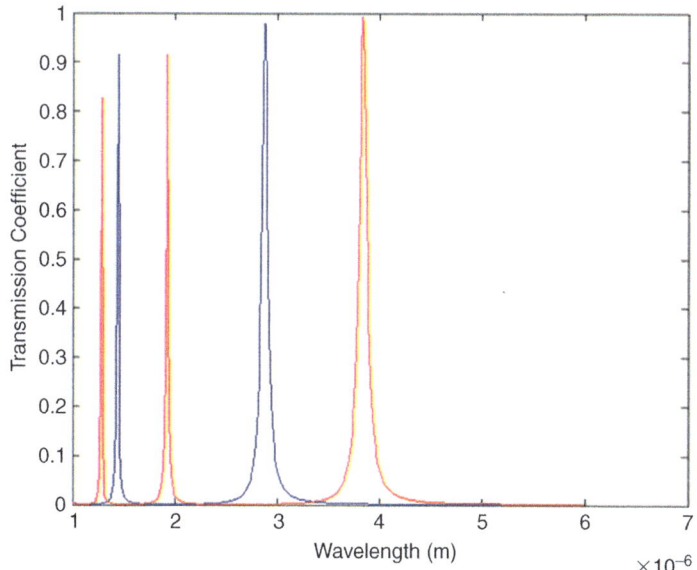

Figure 4.20 Transmission function calculated for the short-wavelength filter using the reflectance data from Figure 4.19. The red plot shows the transmission function for the short wavelength when the air gap is in the relaxed position, and the blue plot shows the transmission function after the air gap has been reduced by 0.5 μm. (Courtesy of Kevin Louchis and Benjamin Hemphill, Team Ninja Star in the Eye, EE115-Winter 2008, Final Report.)

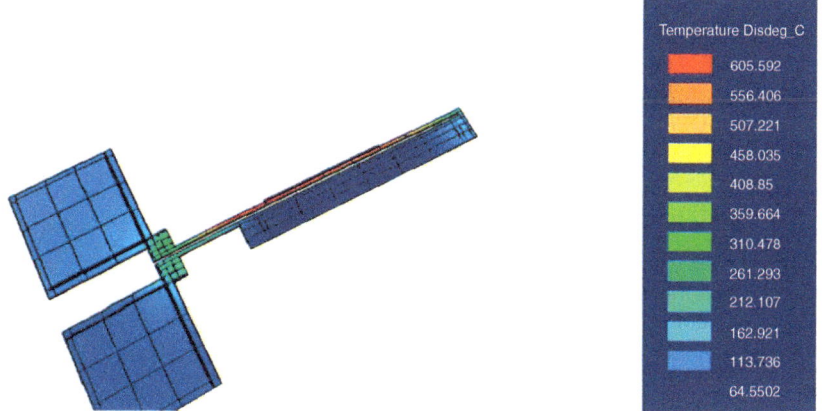

Figure 5.8 Temperature distribution for the heatuator. The hot arm heats up to more than 600°C. (Reprinted with permission from IntelliSuite v8.6 (2010).)

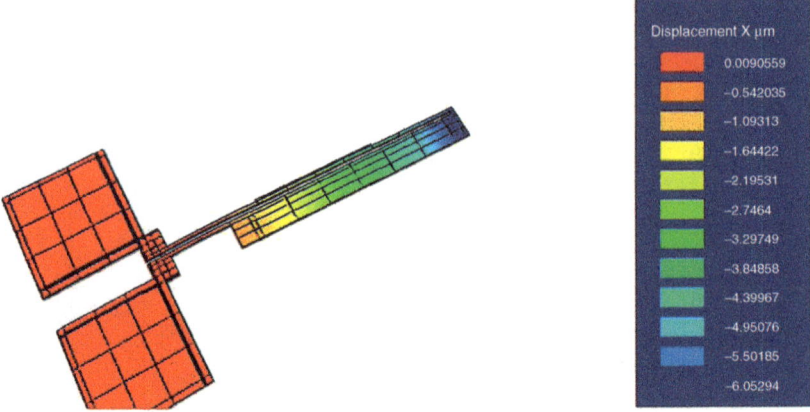

Figure 5.9 Displacement of the heatuator. The tip of the heatuator deflects by more than 6 μm. (Reprinted with permission from IntelliSuite v8.6 (2010).)

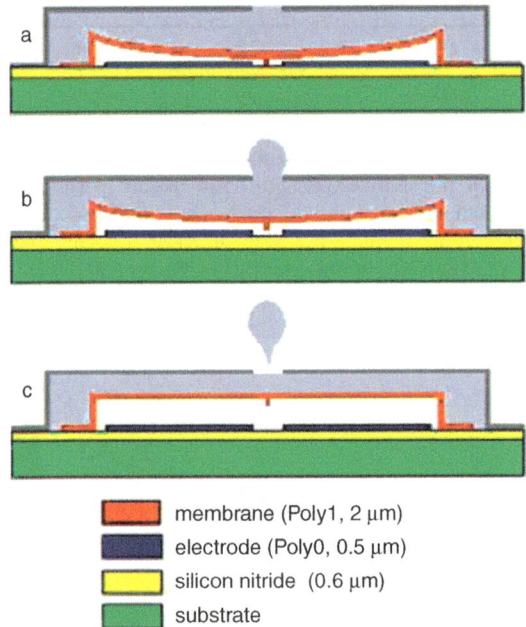

Figure 6.10 Electrostatic membrane drop ejector. (a) The membrane has been pulled down by the application of a voltage applied to the counter-electrode. (b) The voltage is removed and the membrane relaxes back to its initial position, ejecting a drop of ink through the orifice in the nozzle plate, as shown in (c). The membrane is fabricated in Poly1 and the counter-electrode is fabricated in Poly0. The nozzle plate with the orifice can be fabricated in Poly2 or in a separate thick polymer layer such as SU-8 (Xerox MEMSJet).

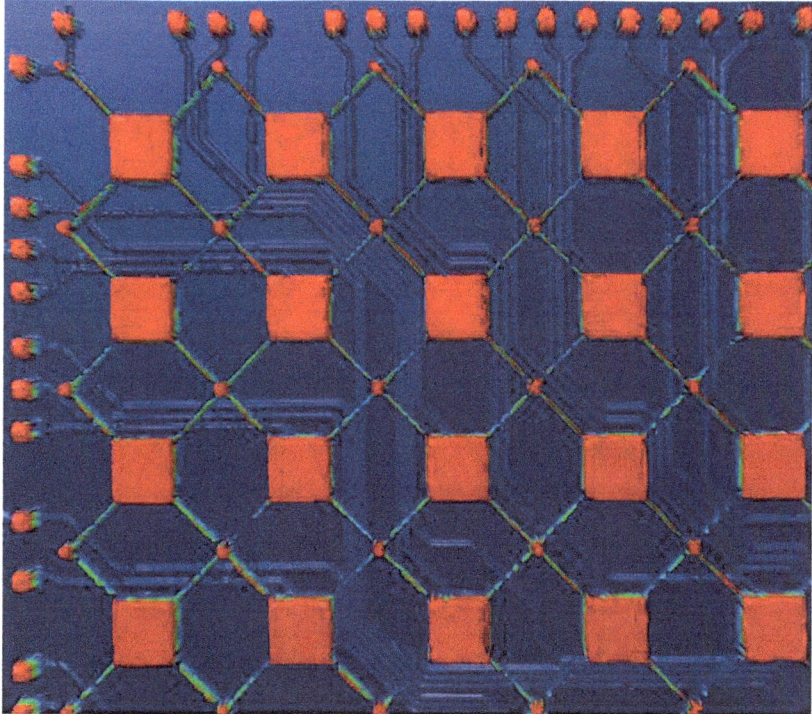

Figure 7.2 A white light interferometer can be used to make measurements of the device under test without contacting it. This technique is most often used to obtain height information for MEMS structures. It can be useful to measure out-of-plane displacement information for vertical actuators.

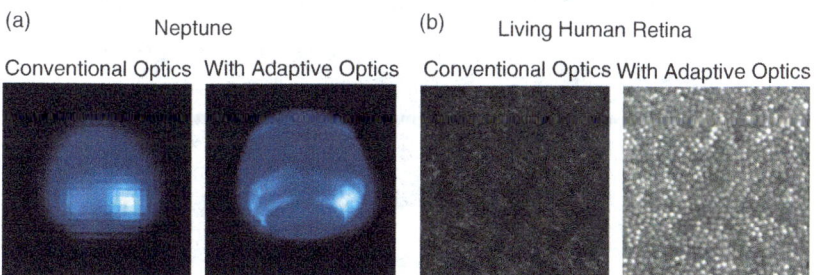

Figure 8.2 The use of adaptive optics to correct image aberrations. (a) Astronomy: Neptune observed in the near-IR (1.65 µm) with and without adaptive optics. (Credit: C.E. Max et al., reproduced by permission of the America Astronomical Society). (b) Vision science: imaging of individual rods and cones in the living human retina. (Credit: Y. Zhang, S. Poonja, and A. Roorda, reproduced by permission of the Optical Society of America.)

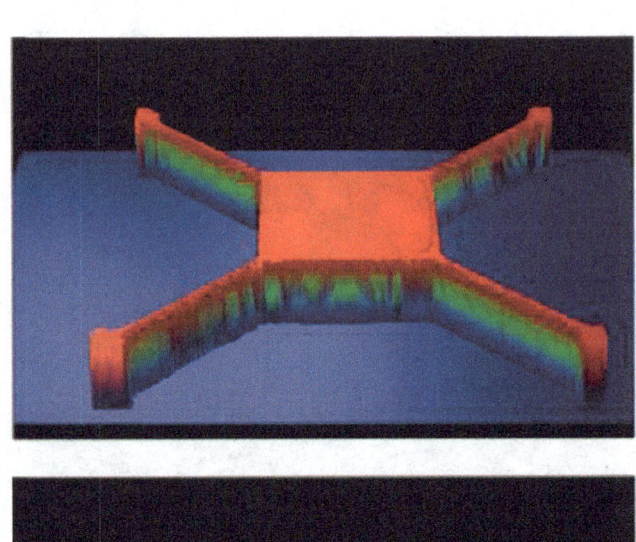
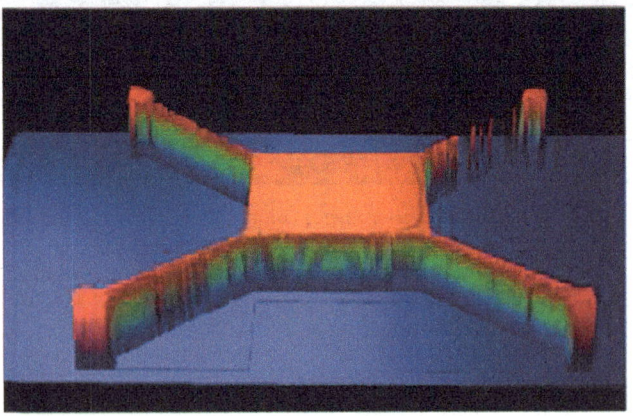
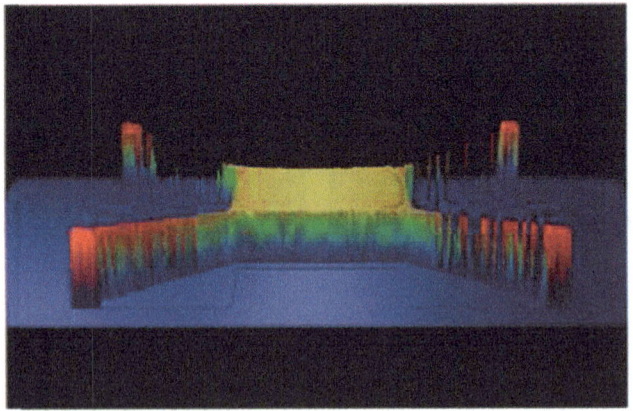

Figure 7.3 Displacement of an individual X-beam actuator. (Top) Undeflected actuator. (Middle) Partial deflection. (Bottom) Maximum deflection.

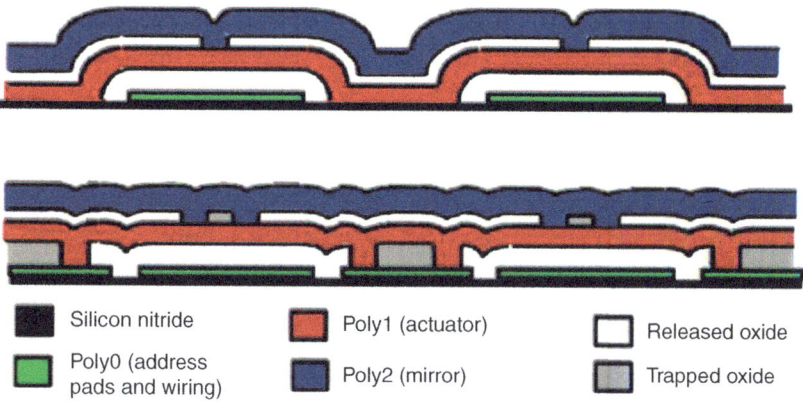

Figure 8.7 The use of trapped oxide to decrease topography due to conformal coatings in the PolyMUMPS process while preserving the strength of the anchors. (Used with permission from Raji Krishnamoorthy Mali, Thomas Bifano, and David Koester, *A design-based approach to planarization in multilayer surface micromachining*, J. Micromechanical Microengineering 9, pp. 294–299 [1999].) [17].

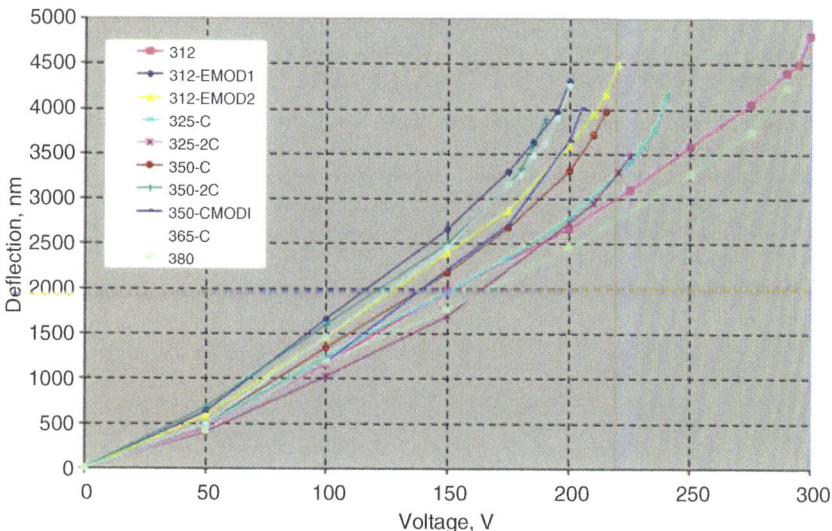

Figure 8.11 Voltage versus deflection characteristics. (Used with permission from S.A. Cornelissen, P.A. Bierden, and T.G. Bifano, *Development of a 4096 element MEMS continuous membrane deformable mirror for high contrast astronomical imaging*, Proc. SPIE 6306, p. 630606-1 [2006].) [21].

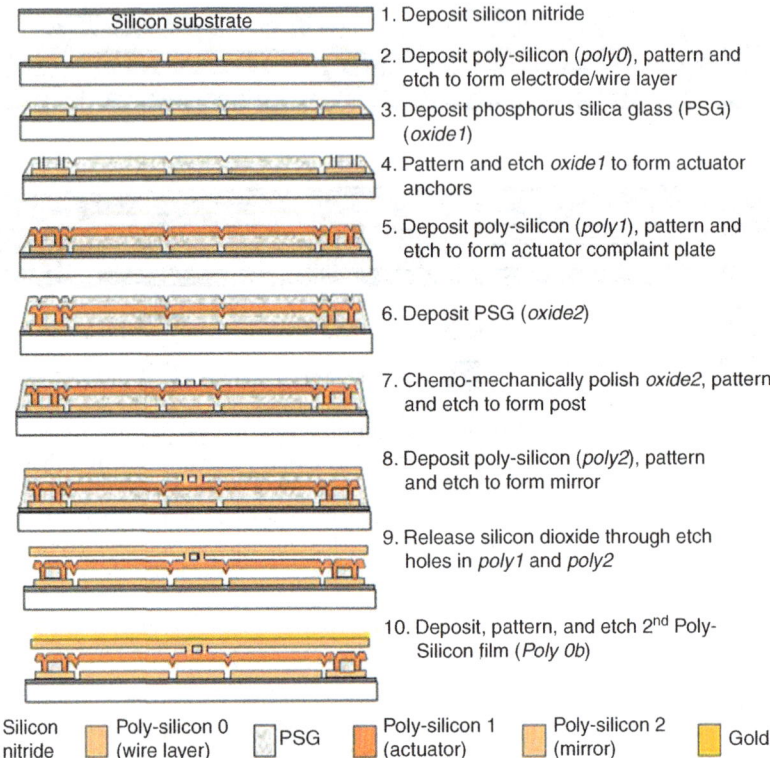

Figure 8.13 Fabrication process flow used to manufacture Boston Micromachines' MEMS DMs. A cross section of a single actuator is shown. (Used with permission from S.A. Cornelissen, P.A. Bierden, and T.G. Bifano, *Development of a 4096 element MEMS continuous membrane deformable mirror for high contrast astronomical imaging*, Proc. SPIE 6306, p. 630606–1 [2006].) [21].

2.11 Pressure sensor

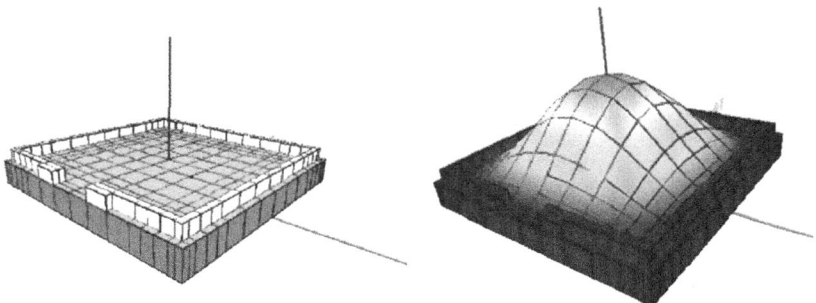

Figure 2.19 Clamped square membrane with a pressure P applied to the back face. (Left) Undeflected membrane. (Right) Deflected membrane with color keyed to displacement. (Reprinted with permission of Sandeep Akkaraju from IntelliSuite v8.6 (2010), IntelliSense Software Corporation.) See color plate section.

$$\begin{aligned}\sigma &= -0.3078\frac{PL^2}{t^2}\\ &= -0.3078\frac{(101\,325\text{ Pa})(100\times 10^{-6}\text{ m})^2}{(1.5\times 10^{-6}\text{ m})^2}\\ &= -1.4\times 10^8\text{ Pa}\\ &= -0.14\text{ GPa}.\end{aligned} \quad (2.46)$$

If we use a piezoresistance of $\pi_{44} = 138 \times 10^{-11}/\text{Pa}$ for p-type polysilicon, then

$$\frac{\Delta R}{R} = \pi_{44}\,\sigma = (138\times 10^{-11}/\text{Pa})(0.14\text{ GPa}) = 0.19. \quad (2.47)$$

Problems

(1) Lay out a comb-drive resonator in the PolyMUMPS process using L-Edit in MEMS Pro as shown in Figure 2.20. Use the Poly1 layer ($h = 2$ μm) for fabrication of the comb-drives and folded springs. You can construct the resonator using elements from the MEMS Pro cell library. The length of the folded springs will be $L = 150$ μm, the width of the beams will be $W = 2$ μm, and they will be separated by 18 μm. The comb-drive fingers will be 40 μm long and 3 μm wide, with a 3 μm gap between the fingers. The fixed and released fingers should have an undeflected overlap of 20 μm. You should not need to use Poly2 in your layout except for the bond pads. Be sure to use bond pads from the MEMS Pro cell library! Save the final

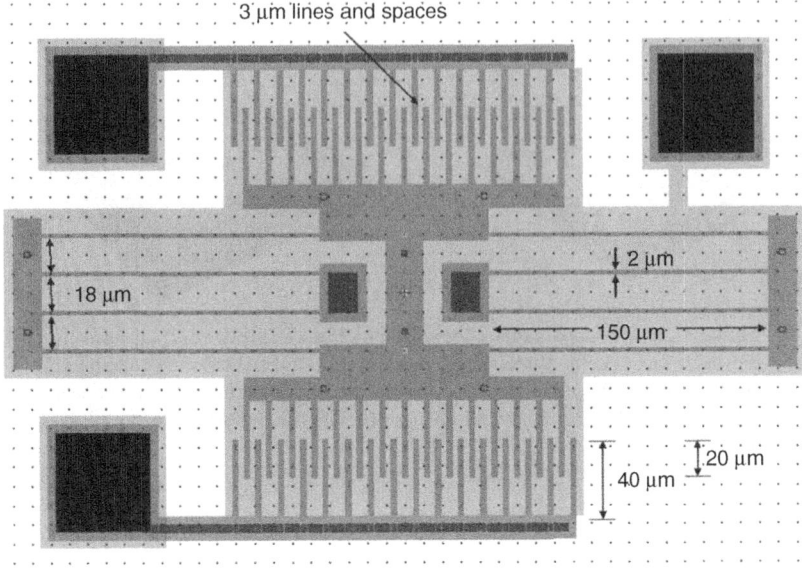

Figure 2.20 Comb-drive resonator.

layout file as Your-Name_Resonator.tdb and mail it to the instructor. Make a solid model of your resonator. Experiment with the scaling in the z-direction to make the solid model easier to visualize.

(2) Calculate the resonant frequency for the comb-drive resonator in Problem 1 above.

(a) First find the effective spring constant k_{sys} for the two folded springs, where each folded spring is made up of four beams of length L (i.e., $L = 150$ μm in the layout in Problem 1). Assume the trusses joining the folded spring segments are rigid, that the Young's modulus for polysilicon is $E = 160$ GPa, and that the density of polysilicon is $\rho = 2330$ kg/m^3. Hint: A folded spring can be broken down into fixed-guided springs that are connected in parallel and in series. Springs that are connected in parallel have an effective spring constant that is the sum of the individual spring constants, $k_{total} = k_1 + k_2$, like capacitors that are connected in parallel, where $C_{total} = C_1 + C_2$. Springs that are connected in series have an effective spring constant given by $1/k_{total} = 1/k_1 + 1/k_2$, like capacitors that are connected in series, where $1/C_{total} = 1/C_1 + 1/C_2$. The spring constant for a single fixed-guided beam is given by

$$k_{fixed-guided} = 12EI/L^3 = Eh(W/L)^3, \quad \text{where } I = (1/12)hW^3.$$

(b) Use the effective spring constant k_{sys} found above to estimate the resonant frequency of the comb-drive resonator using the following formula:

$$f_r = \frac{1}{2\pi}\left[\frac{k_{sys}}{M_p + 0.3714M}\right]^{1/2},$$

where M_P and M are the masses of the shuttle plate and the supporting beams, respectively. This approximate expression is found using the Rayleigh Ritz energy method.

(c) (Extra credit) Find the resonant frequency using finite element analysis (FEM).

(3) Assume that silicon will fracture when the axial stress reaches ≈ 1 GPa. Find the maximum length of a vertical silicon rod that, under the action of its own gravitational load, will not exceed this fracture stress. Assume that the density of silicon is $\rho = 2331$ kg/m^3 and that the acceleration due to gravity is $g = 9.8$ m/s^2 [9].

(4) Explain how you can use the equation for the deflection of a fixed-guided beam subjected to a point load on the guided end to derive the equation for the deflection of a fixed-fixed beam subjected to a point load at its center.

(5) A silicon cantilever of length $L = 500$ μm, width $W = 50$ μm, and thickness $t = 2$ μm is subjected to a uniform *distributed* transverse load ρ [N/m], where $\rho = F/L$. Find the tip deflection $y(L)$ at the end of the cantilever [9]. Solve the equation for the force F and express the solution in the form of Hook's Law, $F = k_{eff}\, y(L)$, with an effective spring constant k_{eff}. What is the effective spring constant k_{eff}? You can leave your answer in terms of E, I, and L. For a load that produces a tip deflection of 2 μm, calculate the maximum stress at the support. Assume the Young's modulus $E = 160$ GPa.

(6) Produce a solid model of a fixed-free cantilever beam in the device layer of the SOIMUMPS process using L-Edit/MEMS Pro or a similar layout tool. An example of a cross section of a cantilever beam (not to scale) is shown in Figure 1.1 of the SOIMUMPS Design Handbook. If the thickness t of the beam is x μm, the width W should be $10x$ μm and the length L should be $100x$ μm. What are

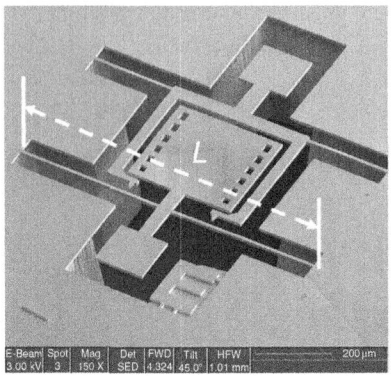 

Figure 2.21 Torsion rods for fold-up mirror in SOIMUMPS process. (Reprinted with permission from Prof. Ash Parameswaran, School of Engineering Science, Simon Fraser University, Burnaby, BC, Canada.)

the allowed values for the thickness of the device layer in the SOIMUMPS process? Derive an expression for the deflection of the beam as a function of a point force F acting on the free end of the beam.

(7) Design and lay out a fold-up mirror in the SOIMUMPS that is rotated out of plane on torsion rods as shown in Figure 2.21. If the torsion rods have a square cross section and are fabricated in a 10 μm thick device layer of an SOI wafer, find the total torsional constant k_θ if each torsion rod is L long. How much "stiffer" would torsion rods of the same length be if they were fabricated with a square cross section in a 25 μm device layer?

(8) Model a released mirror for a MEMS Fabry-Perot interferometer in the SOIMUMPS process as a stiff plate (2 mm × 2 mm) fabricated in the device layer of the 10 μm thick SOI wafer suspended by four diagonal fixed-guided cantilever beams ($h = 10$ μm thick, $W = 100$ μm wide, L long) at each corner of the mirror, as shown in the layout in Figure 2.22. Find an expression for the deflection δ of the mirror as a function of the support arm length L assuming a force F is acting vertically on the mirror. What is the longest length L for the support arms that you can fit into the allowed 8 mm × 8 mm die site including the central 2 mm × 2 mm mirror? You can assume the Young's modulus for silicon is $E = 169$ GPa parallel to the wafer flat for a (100) wafer and 130 GPa at 45 degrees to the wafer flat, and that Poisson's ratio is $v = 0.28$.

2.11 Pressure sensor

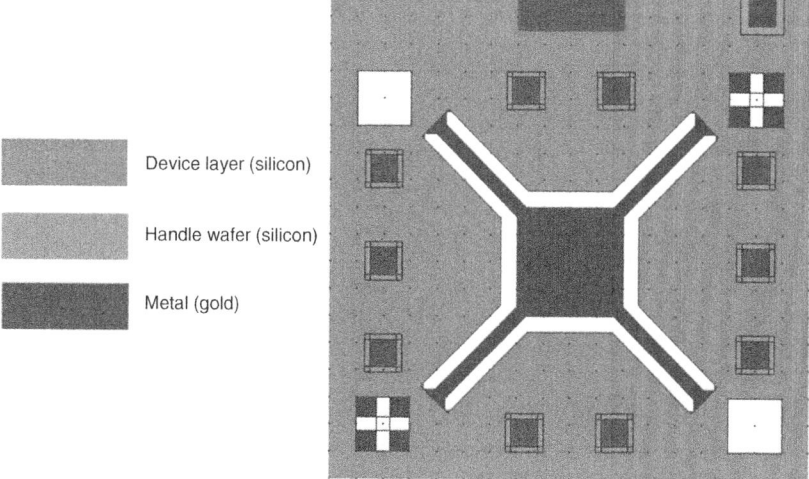

Figure 2.22 Layout for a Fabry-Perot interferometer in the SOIMUMPS process. (From Mr. Dmitry Kozak and the MEMS Design students in EE115, Spring 2009.) See colour plate section.

REFERENCES

1. http://scienceworld.wolfram.com/physics/SpringsTwoSpringsinParallel.html (last accessed on 3/18/2010).
2. http://scienceworld.wolfram.com/physics/SpringsTwoSpringsinParallel.html (last accessed on 3/18/2010).
3. http://en.wikipedia.org/wiki/Image:Poisson_ratio_compression_example.svg (last accessed on 3/18/2010).
4. W.C. Young, *Roark's Formulas for Stress and Strain*, McGraw-Hill (1988).
5. P.M. Osterberg and S.D. Senturia, *M-TEST: A test chip for MEMS material property measurement using electrostatically actuated test structures*, J. Microelectromechanical Systems 6, pp. 107–118 (1997).
6. R.K. Gupta, *Electrostatic pull-in test structure design for mechanical property characterization of microelectromechanical systems (MEMS)*, Ph.D. dissertation, Massachusetts Institute of Technology, Cambridge, MA, 1997.
7. B. Stark, Ed., *MEMS Reliability Assurance Guidelines for Space Applications*, Jet Propulsion Laboratory, Pasadena, California, JPL Publication 99-1, p. 198 (1999).
8. Y.B. Gianchandani and K. Najafi, *Bent-beam strain sensors*, J. Microelectromechanical Systems 5, pp. 52–58 (1996).
9. S.D. Senturia, *Microsystems Design*, Kluwer, p. 238 (2001).

3
Electrostatic Actuation

Electrostatic actuators are commonly used in MEMS devices because they scale well in the micro domain, use very little power, and are straightforward to fabricate in a number of different processes. Two common forms are parallel plate actuators and comb-drive actuators. The parallel plate actuator is a parallel plate capacitor with one of the plates released so that it is able to move, as shown in Figure 3.1. The relationship between the capacitance C, voltage V, and charge Q for a parallel plate capacitor is given by

$$C = \frac{Q}{V}, \qquad (3.1)$$

where the capacitance C is given by

$$C = \varepsilon_0 \frac{A}{g} = \varepsilon_0 \frac{A}{g_0 - z}. \qquad (3.2)$$

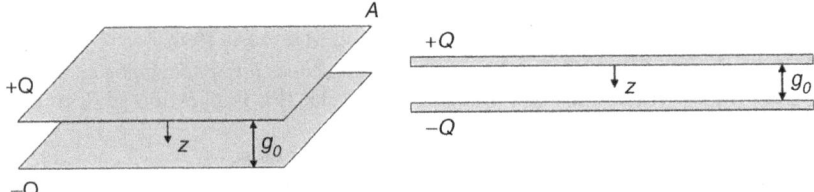

Figure 3.1 Parallel plate capacitor with an area A, charge Q, and an initial gap g_0. When connected to a voltage source, one plate acquires a negative charge ($-Q$) and the other plate acquires a positive charge ($+Q$), leading to an attractive force between the plates. The released plate moves in the z-direction as the gap is decreased from its initial value of g_0.

Electrostatic actuation 59

ε_0 is the dielectric permittivity of free space, 8.85×10^{-12} F/m, g is the distance between the plates (m), and A is the area of the plates (m²).

The incremental work dU done in charging the capacitor by transferring an incremental charge dQ from one plate to the other through a voltage V is given by

$$dU = VdQ. \tag{3.3}$$

Substituting for V,

$$dU = VdQ = \frac{QdQ}{C}. \tag{3.4}$$

Integrating dU for the total work U,

$$U = \int \frac{QdQ}{C} = \frac{1}{2}\frac{Q^2}{C} = \frac{1}{2}CV^2. \tag{3.5}$$

To find the force generated by a parallel plate actuator, we can use the principle of virtual work by considering the work done when the plates of the capacitor are moved a small distance Δz further apart when a constant voltage V, set by a battery, is applied between the plates. The change in the gap causes an amount of charge $\Delta Q = V\Delta C$ to be to be transferred at a potential V from the battery to the capacitor, changing the capacitors stored potential energy. We can then balance the change in the potential energy in the capacitor ($\Delta U_{Capacitor}$) with the mechanical work done to move the plates apart ($\Delta W_{Mechancial}$) and the electrical work done by the battery ($\Delta W_{Battery}$) in transferring the charge ΔQ to maintain the potential at V:

$$\Delta U_{Capacitor} = \Delta W_{Mechancial} + \Delta U_{Battery} \tag{3.6}$$

$$\frac{1}{2}V^2\Delta C = F\Delta z + V\Delta Q \tag{3.7}$$

Using equation (3.1) to substitute for ΔQ at constant V,

$$Q = CV \rightarrow \Delta Q = V\Delta C|_V \rightarrow V\Delta Q = V^2\Delta C \tag{3.8}$$

$$\frac{1}{2}V^2\Delta C = F\Delta z + V^2\Delta C \tag{3.9}$$

$$F\Delta z = -\frac{1}{2}V^2\Delta C \tag{3.10}$$

$$F = -\frac{1}{2}V^2\frac{\Delta C}{\Delta z} \tag{3.11}$$

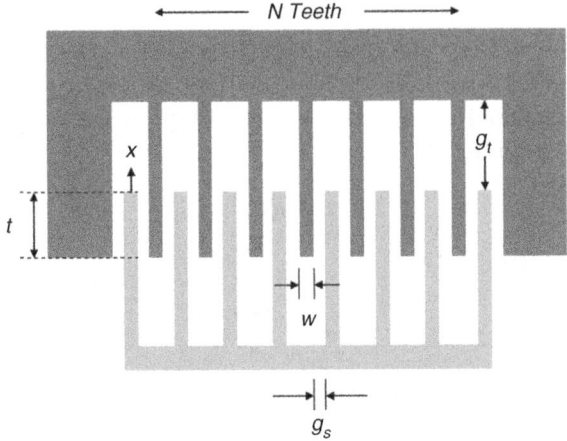

Figure 3.2 Comb-drive actuator. One set of comb teeth is anchored to the substrate and the other set of teeth is released. The N teeth are w wide and have a separation of g_s. The initial overlap between the teeth is given by t, and the initial gap between the anchored and released teeth is given by g_t. When a voltage is applied between the two sets of teeth there is an attractive force that draws the teeth together. The deflection of the teeth is given by x.

We can calculate the force by taking the derivative of the capacitance with respect to the separation between the plates:

$$\left.\frac{\partial C}{\partial z}\right|_V = \frac{\partial}{\partial z}\left(\varepsilon_0 \frac{A}{g_0 - z}\right) = -\varepsilon_0 \frac{A}{(g_0 - z)^2} \quad (3.12)$$

so that the electrostatic force is given by

$$F_e = -\frac{1}{2}V^2 \frac{\Delta C}{\Delta z} = \frac{\varepsilon_0 A}{2} \frac{V^2}{(g_0 - z)^2} \quad (3.13)$$

Since the plate area A scales as the second power of its dimensions and the gap scales as one of the second powers of its dimensions, the electrostatic force in a parallel plate actuator does not scale down with decreasing dimension. For a gap of 1 μm and a voltage of 10 V, the electrostatic force would be 0.44 nN for each square micron of capacitor plate area.

In contrast to the parallel plate actuator, the comb-drive actuator varies capacitance through a change in the overlap area between a set of interpenetrating comb fingers, as shown in Figure 3.2 [1].

Here the initial overlap between the two sets of N fingers is t, the sideways gap between the fingers is g_s, and the gap in the tangential direction of motion is g_t. If the thickness of the film defining the teeth is h, then the capacitance between the two sets of fingers is given by

$$C_s = 2N \frac{\varepsilon_0 h(t+x)}{g_s} \quad (3.14)$$

$$C_t = 2N \frac{\varepsilon_0 h w}{g_t - x}. \quad (3.15)$$

The force that can be developed by the comb-drive actuators is given by

$$F_s = N \frac{\varepsilon_0 h t}{g_s} V^2 \quad (3.16)$$

$$F_t = N \frac{\varepsilon_0 h w}{(g_t - x)^2} V^2 \quad (3.17)$$

The gap between the teeth in the tangential direction of motion g_t, which is determined by layout, is usually designed to be much larger than the spacing between the fingers, g_s, that is determined by the minimum design rules. To increase the total force, an array of N fingers is used.

3.1 Mechanical restoring force

A spring is typically used to apply a mechanical restoring force F_m for electrostatic actuators, as shown in Figure 3.3. The spring can be linear, following Hooke's Law:

$$F_m = -kz, \quad (3.18)$$

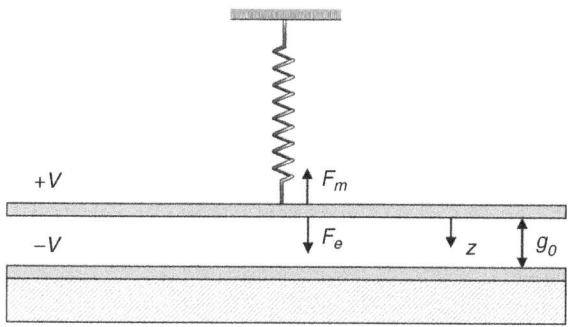

Figure 3.3 Parallel plate electrostatic actuator with a mechanical spring that provides a restoring force F_m in opposition to the attractive electrostatic force F_e. The initial gap is g_0.

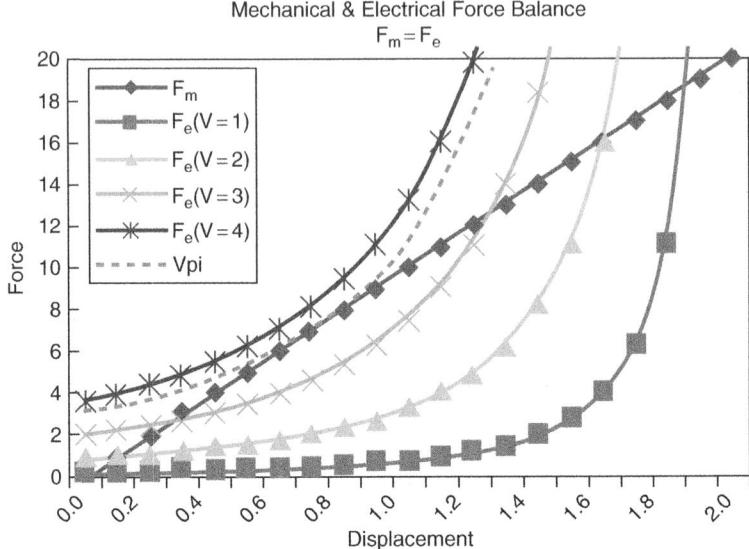

Figure 3.4 Graphical solution for balancing the mechanical and electrical forces. The capacitor has an initial gap g_0 equal to 2.1 μm.

where k is the spring constant and z is the distance the spring is either stretched or compressed. In some situations it can be useful to use a nonlinear spring where the restoring force does not vary linearly with the displacement.

The force balance between the electrostatic force that pulls the released plate down toward the fixed counter electrode and the mechanical restoring force that pulls them apart can be determined graphically as shown in Figure 3.4. The electrostatic force is shown for a few different voltages. There are two solutions for low voltages and no solutions for the highest voltage shown. At a critical voltage, called the "pull-in" voltage, there is only a single solution. The electrostatic force for this single solution is shown as a dashed line. If the voltage is increased further, the nonlinear electrostatic force is greater than the mechanical spring force and the two plates pull in and touch. At the critical voltage, the electrical and mechanical forces are equal:

$$F_m = F_e \qquad (3.19)$$

$$kz = \frac{\varepsilon_0 A}{2} \frac{V^2}{(g_0 - z)^2}. \qquad (3.20)$$

3.1 Mechanical restoring force

The slopes of the electrical and mechanical forces are also equal:

$$\frac{dF_m}{dz} = \frac{dF_e}{dz} \tag{3.21}$$

$$k = \varepsilon_0 A \frac{V^2}{(g_0 - z)^3}. \tag{3.22}$$

Substituting for k and solving for z,

$$\varepsilon_0 A \frac{V^2}{(g_0 - z)^3} z = \frac{\varepsilon_0 A}{2} \frac{V^2}{(g_0 - z)^2} \tag{3.23}$$

$$z = \frac{g_0}{3}. \tag{3.24}$$

When one-third of the initial gap has been closed, the plates snap together or "pull in." This pull-in instability limits the useful range of parallel plate electrostatic actuators with linear springs. For parallel plate electrostatic actuators formed in surface micromachining processes, the initial gap is defined by the sacrificial layer thickness, which is practically limited to a few microns, so that the useful actuation range is typically less than a micron. To find the pull-in voltage, the gap at pull-in, $g_0/3$, can be substituted into equation (3.22) and solved for the voltage:

$$k = \varepsilon_0 A \frac{V^2}{(g_0 - z)^3}\bigg|_{z = \frac{g_0}{3}} = \varepsilon_0 A \frac{V^2}{\left(\frac{2g_0}{3}\right)^3} = \frac{27\varepsilon_0 A}{8g_0^3} V^2 \tag{3.25}$$

$$V_{pull-in} = \sqrt{\frac{8kg_0^3}{27\varepsilon_0 A}}. \tag{3.26}$$

As an example, we consider a parallel plate actuator that is fabricated in the PolyMUMPS process using the first released polysilicon layer, Poly1, as the structural layer and the first oxide, Oxide1, as the sacrificial layer. We consider two configurations for the springs: an X-beam configuration, as shown in Figure 3.5, and a Z-beam configuration, as shown in Figure 3.6.

For the X-beam configuration, the four support springs act in parallel, so the total spring constant K is the sum of the four individual spring constants k. Each spring has fixed-guided boundary conditions. The spring constant for a fixed-guided beam is given by

$$k_{fixed-guided} = \frac{Ewt^3}{l^3}. \tag{3.27}$$

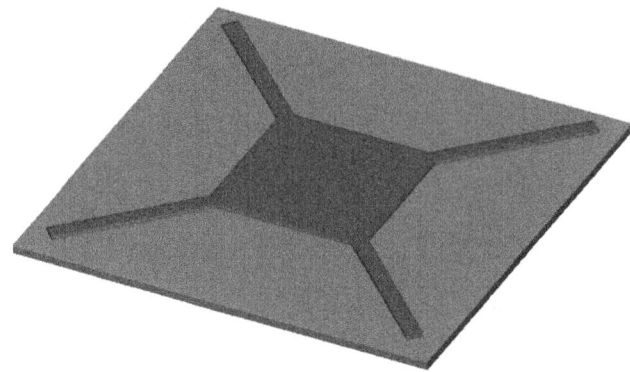

Figure 3.5 X-beam parallel plate actuator fabricated in the PolyMUMPS process.

Figure 3.6 Z-beam parallel plate actuator fabricated in the PolyMUMPS process.

Since the beams are fabricated in Poly1, Young's modulus E would be approximately 160 GPa, and the thickness t would be approximately 2 μm. For support beams that are 10 μm wide and 100 μm long, the spring constant would be $k = 12.8$ N/m, and the total spring constant for all four springs acting in parallel would be $K = 51.2$ N/m.

If we assume that the released plate is 100×100 μm² and approximate it as infinitely stiff because it is 10 times wider than the support beams, we can calculate the pull-in voltage as

$$V_{pull\text{-}in} = \sqrt{\frac{8Kg_0^3}{27\varepsilon_0 A}} = \sqrt{\frac{8(51.2 \text{ N/m})(2 \times 10^{-6} \text{ m})^3}{27(8.85 \times 10^{-12} \text{ F/m})(100 \times 10^{-6} \text{ m})^2}} = 37 \text{ V}. \quad (3.28)$$

Since the pull-in instability occurs at $g_0/3$ for linear springs, the actuator would pull in at approximately 0.7 µm. The useful range could be increased to 0.9 µm by stacking the two sacrificial oxides, Oxide1 (2 µm) and Oxide2 (0.75 µm), and fabricating the released plate in Poly2. We can use scaling to determine the resulting pull-in voltage. Since Poly2 (1.5 µm) is thinner than Poly1 (2.0 µm), the support springs would be more flexible by a factor of $(1.5\,\mu\text{m}/2.0\,\mu\text{m})^3$, so that the pull-in voltage would be decreased by a factor of $(1.5\,\mu\text{m}/2.0\,\mu\text{m})^{3/2}$, to approximately 24V.

The parallel plate actuator can be made more compact by folding the support beams into the side of the actuator plates in the Z-beam configuration shown in Figure 3.6. However, this configuration has a tendency to tilt as it translates downward, so that it suffers from premature pull-in before a displacement of one-third of the gap has been attained [2]. An advantage of the X-beam design is that if one of the arms displaces more than the other arms, it is stretched, which increases its stiffness, providing an increased restoring force. This causes the plate to translate vertically rather than tilting as it is pulled in. Once the arms deflect by more than one arm thickness the stretching also gives rise to a nonlinear spring force, delaying pull-in beyond one-third of the initial gap [3].

3.2 Comb-drive resonator

The comb-drive resonator shown in Figure 3.7 is a common device in surface micromachining processes [4]. We will estimate the resonance frequency by calculating the spring constant of the folded spring and the mass of the released elements that are driven into resonance by electrostatic actuation.

The folded springs are comprised of fixed-guided beams that are connected in parallel and in series, as shown in Figure 3.8.

Each of the fixed-guided support beams would have a spring constant k given by equation (3.27). The total spring constant would be $K = 2k$. If we fabricate the folded spring in Poly1 of the PolyMUMPS process, and if the individual spring elements are $l = 150$ µm long and the minimum feature size $w = 2$ µm wide, the spring constant would be

$$K = 2k_{fixed\text{-}guided} = \frac{2Ewt^3}{l^3} = 2\frac{(160 \times 10^9 \text{ Pa})(2 \times 10^{-6} \text{ m})(2 \times 10^{-6} \text{ m})^3}{(150 \times 10^{-6} \text{ m})^3}$$
$$= 1.5 \text{ N/m}. \tag{3.29}$$

66   *Electrostatic actuation*

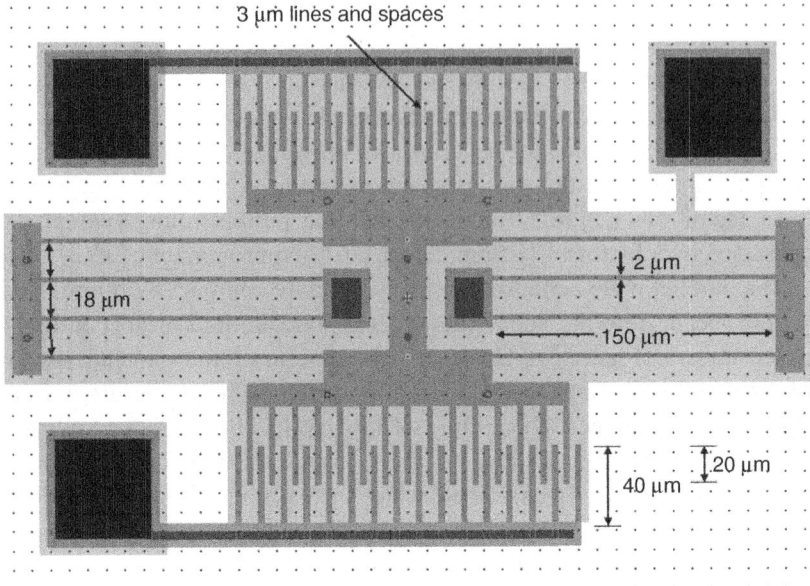

Figure 3.7 Comb-drive resonator.

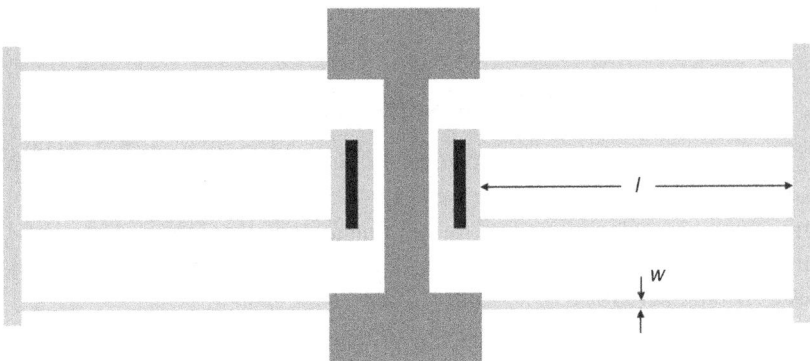

Figure 3.8 Folded spring with fixed-guided beams of length l, width w, and thickness t.

We can estimate the mass of the released elements by considering the shuttle mass and support beams to be concentrated at one point. The mass of the eight polysilicon support beams would be

$$M_{beams} = 8\rho V = 8(2.3 \times 10^3 \, \text{kg/m}^3)(2 \times 10^{-6} \, \text{m})(2 \times 10^{-6} \, \text{m})(150 \times 10^{-6} \, \text{m})$$
$$= 1.1 \times 10^{-11} \, \text{kg}. \tag{3.30}$$

The two $10 \times 150\,\mu m$ connecting beams would have a mass of

$$M_{connectors} = 2\rho V = 2(2.3 \times 10^3 \text{ kg/m}^3)(2 \times 10^{-6} \text{ m})(10 \times 10^{-6} \text{ m})(150 \times 10^{-6} \text{ m})$$
$$= 1.4 \times 10^{-11} \text{ kg}. \tag{3.31}$$

If the shuttle is made up of two plates that are $50 \times 100\,\mu m$ and are connected by a plate that is $50 \times 150\,\mu m$, the mass of the shuttle would be

$$M_{shuttle} = (2.3 \times 10^3 \text{ kg/m}^3)(2(50 \times 10^{-6} \text{ m})(100 \times 10^{-6} \text{ m})(2 \times 10^{-6} \text{ m})$$
$$+ (50 \times 10^{-6} \text{ m})(150 \times 10^{-6} \text{ m})(2 \times 10^{-6} \text{ m}))$$
$$= 8.1 \times 10^{-11} \text{ kg}. \tag{3.32}$$

The total mass would be the sum of the beams, connectors, and shuttles, or 10.6×10^{-11} kg. The resonant frequency can then be approximated by

$$f_R = \frac{1}{2\pi}\sqrt{\frac{k}{m}} = \frac{1}{2\pi}\sqrt{\frac{1.5 \text{ N/m}}{10.6 \times 10^{-11} \text{ kg}}} = 18.9 \text{ kHz}. \tag{3.33}$$

3.3 Cantilever beam resonator

A fixed-free cantilever beam M-Test structure fabricated in Poly1 of the PolyMUMPS process can also be used as a resonator by exciting it into resonance with an AC drive signal applied to a Poly0 counter-electrode defined below the end of the beam. To estimate the resonant frequency we use the spring constant for a fixed-free beam that is $20\,\mu m$ wide and $150\,\mu m$ long subjected to a point load at its free end:

$$k = \frac{Ewt^3}{4l^3} = \frac{(160 \times 10^9 \text{ Pa})(20 \times 10^{-6} \text{ m})(2 \times 10^{-6} \text{ m})^3}{4(150 \times 10^{-6} \text{ m})^3} = 1.9 \text{ N/m} \tag{3.34}$$

$$M_{beam} = \rho V = (2.3 \times 10^3 \text{ kg/m}^3)(2 \times 10^{-6} \text{ m})(20 \times 10^{-6} \text{ m})(150 \times 10^{-6} \text{ m})$$
$$= 1.4 \times 10^{-11} \text{ kg}. \tag{3.35}$$

The resonant frequency would then be approximately

$$f_R = \frac{1}{2\pi}\sqrt{\frac{k}{m}} = \frac{1}{2\pi}\sqrt{\frac{1.9 \text{ N/m}}{1.4 \times 10^{-11} \text{ kg}}} = 58.6 \text{ kHz}. \tag{3.36}$$

3.4 Fixed-fixed beam resonator

The fixed-fixed beam M-Test structure is stiffer than the fixed-free cantilever beam test structure, and would have a higher resonant frequency. Assuming a point load due to a Poly0 counter-electrode below the center of the beam, the spring constant for a fixed-fixed beam that is 20 µm wide and 150 µm long subjected to a point load at its middle would be

$$k = 2\frac{Ewt^3}{l^3} = 2\frac{(160 \times 10^9 \text{ Pa})(20 \times 10^{-6} \text{ m})(2 \times 10^{-6} \text{ m})^3}{(150 \times 10^{-6} \text{ m})^3}$$
$$= 15.2 \text{ N/m}. \tag{3.37}$$

The resonant frequency would then be approximately

$$f_R = \frac{1}{2\pi}\sqrt{\frac{k}{m}} = \frac{1}{2\pi}\sqrt{\frac{15.2 \text{ N/m}}{1.4 \times 10^{-11} \text{ kg}}} = 166 \text{ kHz}. \tag{3.38}$$

Problems

(1) Calculate the voltage that will be required to deflect the mirror for the Fabry-Perot interferometer you designed in Chapter 2 using an electrostatic actuator. You can assume that the area of the parallel plate capacitor is the same as the 2 × 2 mm mirror surface. Assume the initial gap is 2.4 µm and that you need to be able to deflect the mirror surface until the final gap is two-thirds of the initial value, or 1.6 µm.

(2) Explain in words why a parallel plate actuator such as the one shown in Figure 3.3 exhibits a pull-in instability.

(3) Two M-Test structures [5] are shown in Figure 3.9. These test structures are useful for characterizing the mechanical properties of thin films such as Young's modulus, Poisson's ratio, and residual stress. They are also very common elements in MEMS design. In this problem we will determine the pull-in voltage of the cantilever beam (CB) and fixed-fixed beam (FB) analytically.

  (a) Lay out each of these structures in L-Edit. Define the released structure (cantilever beam or fixed-fixed beam, both 500 µm long and 100 µm wide) in Poly1. Instead of the dielectric spacer, as shown in Figure 3.9, use a polysilicon anchor to nitride. Make the ground plane in Poly0. You can start the ground plane after the

3.4 Fixed-beam resonator

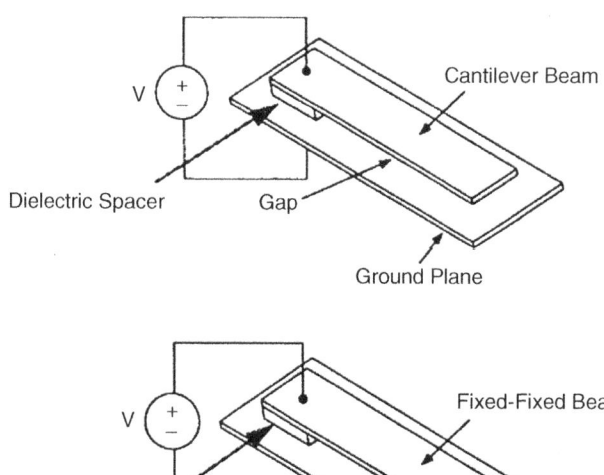

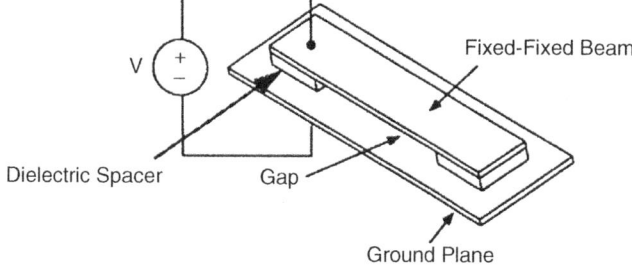

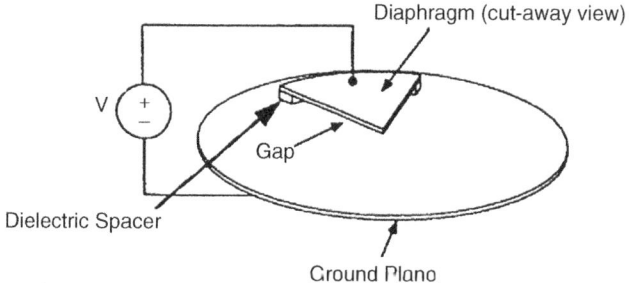

Figure 3.9 M-Test structures. Cantilever beam (top) and fixed-fixed beam (bottom). (Reprinted with permission from J. Microelectromechanical Systems, *M-TEST: a test chip for MEMS material property measurement using electrostatically actuated test structures*, P.M. Osterberg and S.D. Senturia, ©1967 IEEE.)

nitride anchor. Use Oxide1 to define a 2 μm gap between the released structure and the Poly0 ground plane.
(b) Make a solid model of each of these structures using MEMS Pro. Draw representative cross sections through your structures.
(c) Calculate the pull-in voltage for each of these structures analytically for the following dimensions:

Cantilever beam (CB):

	Length	Width
	300 μm	100 μm
	400 μm	100 μm
	500 μm	100 μm

Fixed-fixed beam (FB): Same as above

Assume the following materials properties:
- Young's Modulus $E = 160$ GPa
- Poisson's ratio $v = 0.2$
- Transverse stress gradient $= 0$
- Oxide1 thickness $= 2$ μm
- Poly1 thickness $= 2$ μm

You may also assume that each of these structures can be approximated as an ideal parallel plate actuator, and that there is a uniform electrostatic load on each of them, as shown in Figure 3.10. You can also ignore the length of the beam taken up by the anchors in your calculations, and any fringing effects of the fields. The displacements would then be given by

Cantilever beam:

$$y(x) = \frac{px^2}{24EI}\left(6\ell^2 - 4\ell x + x^2\right).$$

Fixed-fixed beam:

$$y(x) = \frac{px^2}{24EI}(\ell - x)^2,$$

where the moment of inertia I is given by

$$I = \frac{wt^3}{12}.$$

To calculate the pull-in voltages, first determine an effective spring constant k_m for each of the structures. Note that the maximum displacement for the cantilever beam occurs at $x = L$ and that the maximum displacement of the fixed-fixed beam occurs at $x = L/2$. Then calculate their initial capacitance C_0. The pull-in voltage can then be found from equation (4.18) in Liu [6]:

$$V_p = \frac{2x_0}{3}\sqrt{\frac{k_m}{1.5C_0}}.$$

3.4 Fixed-beam resonator

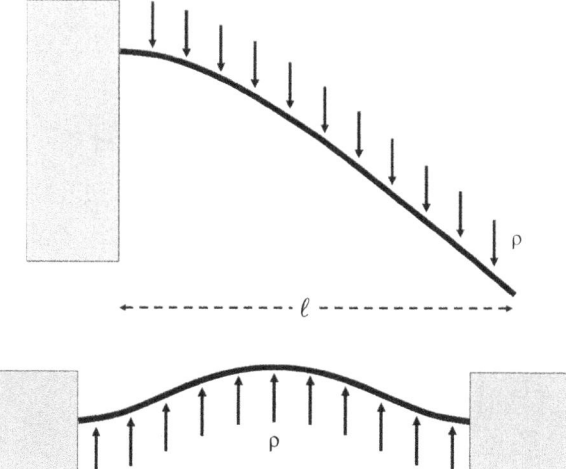

Figure 3.10 Cantilever beam under uniform load $\rho = F/L$ (N/m) (top), fixed-fixed beam (bottom).

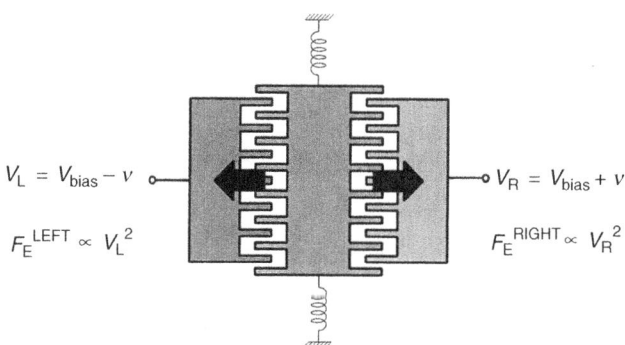

Figure 3.11 Pull-pull comb-drive. (Reprinted with permission from Prof. Hiroshi Toshiyoshi, Institute of Industrial Science (IIS), The University of Tokyo, Japan.)

(4) Show that the electrostatic force F_E of a "pull-pull" comb-drive actuator, as shown schematically in Figure 3.11, can be linearized in voltage by applying $V_L = V_{bias} - v$ to the left set of fixed comb-drive fingers and $V_R = V_{bias} + v$ to the right set of fixed comb-drive fingers.

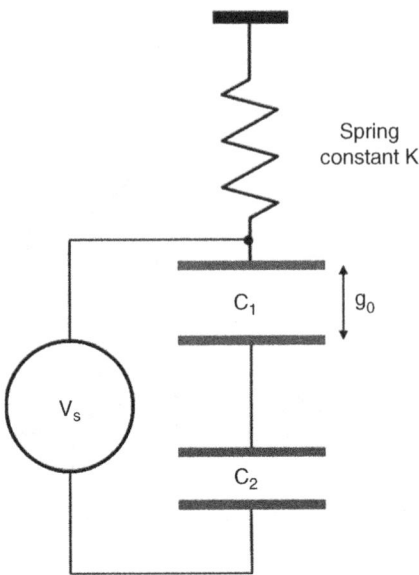

Figure 3.12 Parallel plate electrostatic actuator with a feedback capacitor. The top plate of capacitor C_1 has been released and the bottom plate is fixed in position. The spring provides a mechanical restoring force to the attractive electrostatic force when a voltage V_S is applied. Both the top and bottom plates of capacitor C_2 are fixed.

(5) A parallel plate electrostatic actuator with a variable capacitance C_1 is in series with a fixed feedback capacitance C_2, as shown in Figure 3.12 [7].
  - Find the total capacitance of the system at zero-voltage (e.g., the gap g of the variable capacitor is g_0). Assume both capacitors have the same area A and define $R = C_1(V=0)/C_2$.
  - Find the voltage across the variable capacitor.
  - Find the electrostatic force as a function of the supply voltage.
  - Derive the stability condition (pull-in). Show that the pull-in displacement depends on the value of the feedback capacitance.
  - If we would like to eliminate pull-in completely, what capacitance value (measured in terms of the capacitance value of the parallel plate actuator) should we use?

(6) Find the size of the gap z at pull-in relative to the initial gap g_0 for an electrostatic parallel plate actuator with a nonlinear spring that follows Snook's Law:

$$F_m = -kz^3. \tag{3.39}$$

REFERENCES

1 W.C. Tang, T.-C.H. Nguyen, M.W. Judy, and R.T. Howe, *Electrostatic-comb drive of lateral polysilicon resonators*, Sensors and Actuators A21–23, pp. 328–331 (1990).
2 J.I. Seeger and B.E. Boser, *Charge control of parallel-plate, electrostatic actuators and the tip-in instability*, J. Microelectromechanical Systems 12, pp. 656–671 (2003).
3 E.S. Hung and S.D. Senturia, *Extending the travel range of analog-tuned electrostatic actuators*, J. Microelectromechanical Systems 8, pp. 497–505 (1999).
4 W.C. Tang, T.-C.H. Nguyen, and R.T. Howe, *Laterally driven polysilicon resonant microstructures*, in Proc. IEEE Workshop on Microelectromechanical Systems, Salt Lake City, UT, Feb. 1989, pp. 53–59 (1989).
5 P.M. Osterberg and S.D. Senturia, *M-Test: A test chip for MEMS material property measurement using electrostatically actuated test structures*, J. Microelectromechanical Systems 5, pp. 107–118 (1997).
6 C. Liu, *Foundations of MEMS*, Pearson Prentice Hall, p. 113 (2006).
7 J.I. Seeger and S.B. Crary, *Stabilization of electrostatically actuated mechanical devices*, Transducers'97, 1997 International Conference on Solid-state Sensors and Actuators, Chicago, June 16–19, 1997, pp. 1133–1136.

4
Optical MEMS

Optical MEMS are based on reflection, refraction, diffraction, and interference of light. MEMS technology is well suited for optical applications because light has no mass and so typically only small forces are required for actuation, and light can be passed through an optical window that protects the device from the ambient environment. A number of optical MEMS devices have made it to market including digital projectors, projection TVs, scanners, and displays, both for handheld devices and for head mount displays. Many optical MEMS devices were also developed during the "dot com" bubble, including optical switches, cross-connects, variable optical attenuators, tunable lasers, and tunable filters, although most of these devices never made it to market due to the bursting of the bubble. Some of these development efforts made use of the same multiproject wafer processes in their early prototyping stages and will be examined in detail as case studies here.

4.1 Reflecting cantilever beam optical modulator

One of the earliest optical MEMS devices used the electrostatic deflection of an array of cantilever beams in combination with a galvo scanner to form a projection display system as shown in Figure 4.1 [1]. The cantilever beams were metal-coated silicon dioxide, but they could also be formed in polysilicon using the MUMPS process or in single crystal silicon in the SOIMUMPS process.

The electrostatic deflection of a cantilever beam has been derived by Petersen [2] and Kovacs [3]. The deflection of a cantilever beam of width w and length l due to a point load F acting on the end of a beam was found to be

4.1 Reflecting cantilever beam optical modulator

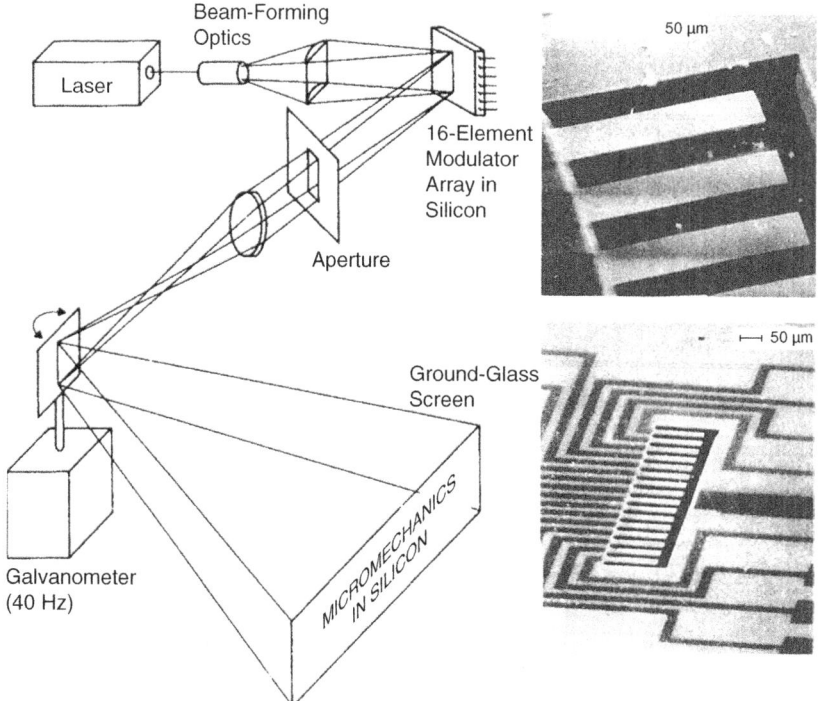

Figure 4.1 MEMS projection display that uses a 16-element array of cantilever beams to modulate light. Light that is reflected off the cantilever beam array into the aperture is scanned across a ground-glass screen by a galvo scanner to form a projection image. (Reprinted with permission from Proc. IEEE, *Silicon as a mechanical material*, Kurt E. Petersen, ©1967 IEEE.)

$$y(x) = \frac{Fx^2}{6EI}(3l - x) \quad I = \frac{wt^3}{12}. \tag{4.1}$$

If we consider a normalized point force $q(x) = F(x)/A$ acting on a small segment of the cantilever beam wdx at the position x,

$$dy = \frac{x^2}{6EI}(3l - x)dF = \frac{x^2}{6EI}(3l - x)q(x)wdx. \tag{4.2}$$

For an electrostatic force on the cantilever with an initial gap d_0, the force would be

$$F(x) = \frac{\varepsilon_0 A}{2}\frac{V^2}{(d_0 - d(x))^2} \rightarrow q(x) = \frac{F(x)}{A} = \frac{\varepsilon_0}{2}\frac{V^2}{(d_0 - d(x))^2}. \tag{4.3}$$

The total tip deflection can be found from integrating the incremental deflection along the length of the cantilever beam:

$$y(l) = \int_0^l \frac{x^2}{6EI}(3l-x)q(x)w\,dx = \frac{\varepsilon_0 w V^2}{12EI}\int_0^l \frac{x^2(3l-x)}{(d_0-d(x))^2}\,dx. \qquad (4.4)$$

To solve this equation in closed form, Kovacs made a parabolic approximation for the gap $d(x)$ as a function of the tip deflection $y(l)$:

$$d(x) \approx \left(\frac{x}{l}\right)^2 y(l) \qquad (4.5)$$

$$y(l) \approx \frac{\varepsilon_0 w V^2}{12EI}\int_0^l \frac{x^2(3l-x)}{\left(d_0 - \frac{xy(l)}{l}\right)^2}\,dx, \qquad (4.6)$$

which can be solved for $y(l)$. To find the pull-in voltage V_{pi}, this equation can be solved for $y(l) = d_0$.

The problem can also be solved using coupled electromechanical modeling in IntelliSuite with the deflection of the cantilever beam shown in Figure 4.2.

Another possibility for driving the cantilever beam light modulator would be to form a bimorph that is heated to cause it to deflect. A gold

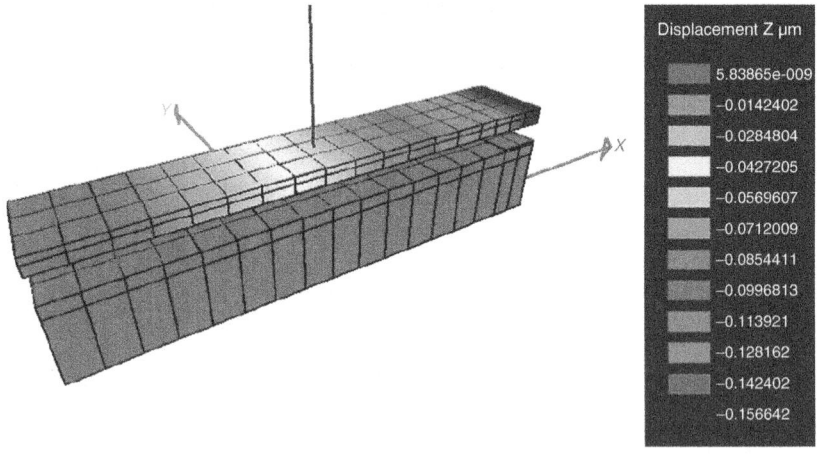

Figure 4.2 Deflection of an electrostatically actuated cantilever beam solved in IntelliSuite. (Reprinted with permission of Sandeep Akkaraju from IntelliSuite v8.6, 2010, IntelliSense Software Corporation.) See color plate section.

coating on a silicon or polysilicon beam would provide the required CTE mismatch and reflectivity, so long as it is kept below the gold/silicon eutectic temperature.

4.2 Single-axis torsional mirror

A single-axis galvo scanner can be fabricated in the SOIMUMPS by forming the mirror in the device layer of the SOI wafer and bonding the SOI wafer to a glass wafer that has counter-electrodes to form an electrostatic actuator (see Figure 4.3). By using a SOI wafer a thick mirror can be obtained that is useful to minimize static mirror deformations when it is metallized and dynamic mirror deformations when it is actuated. The SOIMUMPS process offers SOI device layers thicknesses of 10 μm or 25 μm. A schematic diagram of a tip-tilt mirror used for a laser scanner formed in a silicon wafer and bonded to a glass wafer that has counter-electrodes is shown in Figure 4.4 [4], [5], [6]. The CTE of the glass wafer (Corning 7740) was matched to the CTE of the silicon mirror wafer. A 50 μm thick resist was used as a spacer layer to form the gap between the mirror and the counter-electrodes.

The deflection angle θ_0 of a single-axis torsional mirror as a function of applied voltage V has been solved analytically by Senturia [7] in cylindrical coordinates, which we follow here. The mirror is grounded and a voltage V is applied to the counter-electrode. The mirror tilt angle θ is positive in the clockwise direction (Figure 4.5).

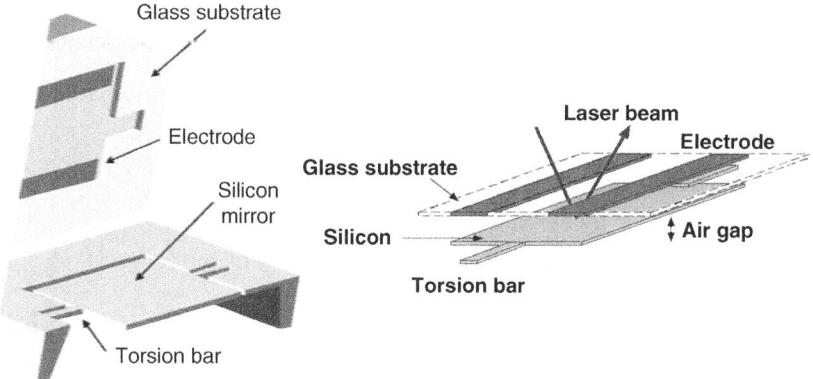

Figure 4.3 Schematic diagram of a torsional mirror used for a laser scanner. (Reprinted with permission of Eric Peeters, PARC.)

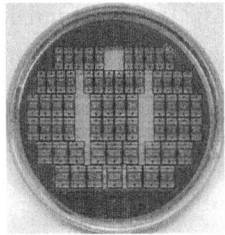 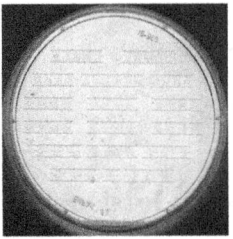 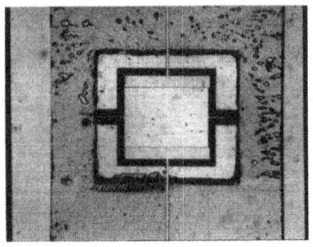

Figure 4.4 Silicon torsional mirror for a laser scanner. (Left) Glass wafer (Corning 7740) with counter-electrodes and 50 μm thick polyimide spacer layer. (Middle) Silicon mirror wafer with aluminum metallization. (Right) Bonded glass and silicon wafers with polyimide spacer layer.

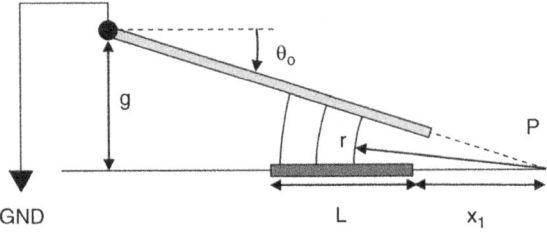

Figure 4.5 A torsional mirror is tilted at an angle θ_0 when a voltage V is applied to a counter-electrode of length L and width W. The mirror is grounded. The initial gap between the mirror and the counter-electrode is g. The projection of the mirror intersects the axis at a point P.

The potential $\phi(\theta)$ of the mirror is given by

$$\phi(\theta) = V\left(1 - \frac{\theta}{\theta_0}\right). \qquad (4.7)$$

The electric field can be found from the gradient of the potential. In cylindrical coordinates, the gradient is given by

$$\nabla \equiv \hat{r}\frac{\partial}{\partial r} + \hat{\theta}\frac{1}{r}\frac{\partial}{\partial \theta} + \hat{z}\frac{\partial}{\partial z}, \qquad (4.8)$$

and the electric field E is found from the gradient of the potential:

$$E = -\nabla\phi = -\hat{\theta}\frac{1}{r}\frac{\partial}{\partial \theta}V\left(1 - \frac{\theta}{\theta_0}\right) = \hat{\theta}\frac{V}{r\theta_0}. \qquad (4.9)$$

The charge density on the plate on the counter-electrode is given by $\varepsilon_0 E$, so that the total charge on the counter electrode is given by

4.2 Single-axis torsional mirror

$$Q = \frac{\varepsilon_0 WV}{r\theta_0} \int_{P-(x_1+L)}^{P-x_1} \frac{dr}{r} = \frac{\varepsilon_0 WV}{r\theta_0} \ln(r)\Big|_{P-(x_1+L)}^{P-x_1} = \frac{\varepsilon_0 WV}{r\theta_0} \ln\left(\frac{P-x_1}{P-(x_1+L)}\right), \quad (4.10)$$

and the capacitance can be found from

$$C = \frac{Q}{V} = \frac{\varepsilon_0 W}{r\theta_0} \ln\left(\frac{P-x_1}{P-(x_1+L)}\right), \quad (4.11)$$

where

$$P = g\cot(\theta_0). \quad (4.12)$$

The torque on the mirror can be found from the principle of virtual work as described in Chapter 3, equation (3.6). Here, if the mirror angle θ_0 increases a small amount $\Delta\theta$, the mechanical work done by the electrostatic actuator will be

$$\Delta W_m = \tau_e \Delta\theta_0. \quad (4.13)$$

The force between the mirror and the counter-electrode is attractive, so that work is done by the mirror when the distance between them decreases. The work done by the mirror plus the work done by the battery must be equal to the change in the potential energy of the mirror. If the voltage on the actuator is held constant, as the mirror rotates toward the counter-electrode, the battery will have to supply charge ΔQ to keep the voltage constant as the capacitance increases, where ΔQ is give by

$$C = \frac{Q}{V} \to Q = CV \to \Delta Q = V\Delta C. \quad (4.14)$$

The battery must do work ΔW_e to supply this charge:

$$\Delta W_e = V\Delta Q = V^2\Delta C. \quad (4.15)$$

The work done by the actuator plus the work done by the battery must be equal to the change in the potential energy of the mirror:

$$\Delta W_{Capacitor} + \Delta W_{Battery} = \Delta U_{Capacitor} \quad (4.16)$$

$$\tau_e \Delta\theta + V^2\Delta C = \frac{1}{2}V^2\Delta C \to \tau_e = -\frac{1}{2}V^2\frac{dC}{d\theta_0}. \quad (4.17)$$

This equation can be solved for the electrostatic torque τ_e by differentiating the capacitance with respect to the mirror tilt angle. The mechanical restoring torque τ_m is given by the two torsion rods that support the mirror. From equation (2.20), for a single torsion rod of length L, torsional moment K, and shear modulus G,

$$\theta_0 = \frac{\tau L}{KG} \rightarrow \tau = \frac{KG}{L}\theta_0 = k_\theta \theta_0, \qquad (4.18)$$

where the torsional spring constant k_θ is given by

$$k_\theta = \frac{KG}{L}. \qquad (4.19)$$

Since the mirror is supported by two torsion rods, the total mechanical restoring torque τ_m would be $\tau_m = 2\tau$. The tilt angle θ_0 as a function of voltage can be found from equating the electrostatic torque τ_e to the total mechanical restoring torque τ_m. The pull-in voltage can be found from setting the torques and the slopes of the torques equal, as was done for the parallel plate actuator:

$$\tau_e = \tau_m \qquad (4.20)$$

$$\frac{d\tau_e}{d\theta_0} = \frac{d\tau_m}{d\theta_0}. \qquad (4.21)$$

A MEMS torsional mirror has also been modeled using coupled electromechanical modeling as shown in Figures 4.6 and 4.7.

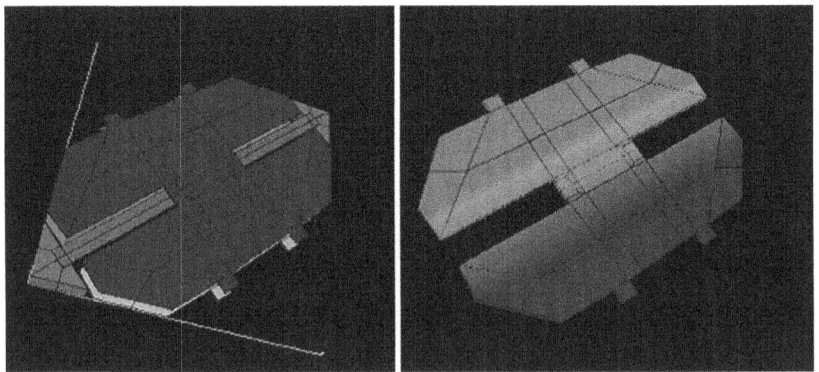

Figure 4.6 Coupled electromechanical modeling of a MEMS torsional mirror. (Reprinted with permission of Sandeep Akkaraju from IntelliSuite v8.6 (2010), IntelliSense Software Corporation.) See color plate section.

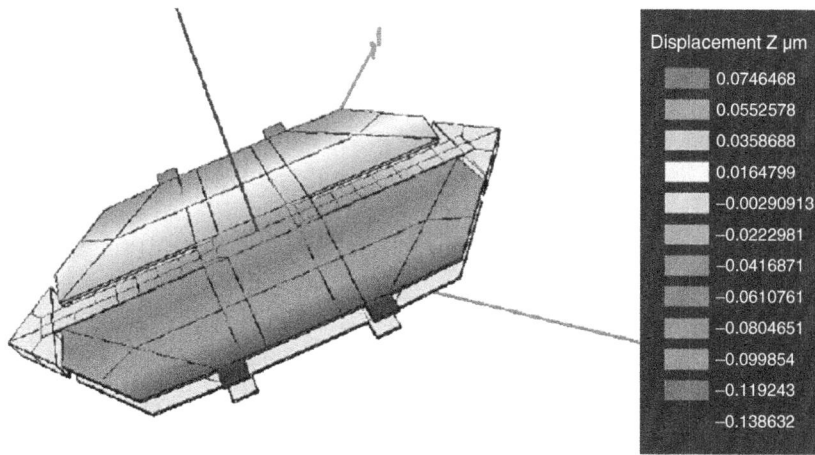

Figure 4.7 Displacement for a MEMS torsional mirror. (Reprinted with permission of Sandeep Akkaraju from IntelliSuite v8.6, 2010, IntelliSense Software Corporation.) See color plate section.

4.3 Dual-axis torsional mirror: Lucent lambda router optical switch

Two-axis beam steering MEMS mirrors were developed by Lucent for switching optical signals in telecommunications [8], [9], [10]. Rather than converting the signal in the optical domain into the electrical domain for switching, as has been done in the past, it can be kept in the optical domain to decrease the cost of the switch. This eliminates the need for a detector to convert the optical signal into an electrical signal, and for a laser to convert the electrical signal back into an optical signal after it has been switched in the electrical domain. The electro-optical conversion gets very expensive for wavelength division multiplexed signals, where many signals of different wavelengths are combined, each needing a dedicated laser of the right color for the electro-optic conversion. In addition, if the signal is converted into the electrical domain, the electronics must be upgraded as the data rate increases. If the signals are left in the optical domain, switching is independent of data rate, format, and the number of multiplexed signals.

Two different types of optical switches have been developed. One is called 3-d, for three-dimensional switching, where optical signals are directed through a 3-d space. The other is called 2-d, because the optical signals are directed in the two-dimensional space above a wafer surface. For performance reasons, 3-d switches are used in larger arrays ($n > 32$),

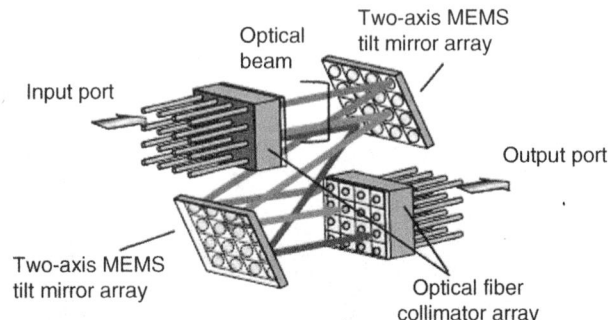

Figure 4.8 Schematic diagram of a 3-d optical cross-connect switch. Light from an input port is reflected off a two-axis MEMS tilt mirror array. The reflected optical beams are then directed to a second two-axis MEMS tilt mirror array that directs the beams into an optical fiber collimator at the output port. (Reprinted with permission from IEEE Conference Proceedings on Photonics in Switching, PS '09, *128 × 128 3D-MEMS optical switch module with simultaneous optical paths connection for optical cross-connect systems*, M. Mizukami et al. ©1967 IEEE.)

and 2-d switches are used in arrays with 32 channels or less. We will consider both types of switches.

The Lucent lambda router, a 3-d switch, was developed as an optical cross-connect switch for large switch arrays (256×256 to 1024×1024 port counts). A schematic diagram of a 3-d optical cross-connect switch is shown in Figure 4.8 [11].

In choosing a process for fabricating the mirror array the developers used the PolyMUMPS process, which allowed them to focus on the design of the mirror rather than on developing a new process. By designing into an existing process they were able to have a very fast turnaround time for fabrication, allowing them to fully optimize the design and fine tune the essential process parameters. The maturity of the process allowed them to exceed performance specifications and attain high yield, which was important for the fabrication of large arrays of mirrors. To overcome some of the shortfalls of the established process, as we discuss later, they had to develop an innovative approach for out-of-plane self assembly of the mirror array.

A picture of the surface micromachined beam steering mirror is shown in Figure 4.9 [10]. The central mirror is 500 μm in diameter and is suspended from a fixed frame by a gimbal mount. The fixed frame has been lifted out of plane by hinged sidewalls, which are locked into place by dovetails in the frame that fit into tapered cuts in the sidewalls. Lifting the frame out of plane overcomes the limited electrostatic gap that can be

4.3 Dual-axis torsional mirror

Figure 4.9 Surface micromachined beam steering mirror. The circular mirror is 500 μm in diameter and is attached to the gimbal ring by two serpentine torsional springs that allow for tip-tilt motion along a first axis. The gimbal ring is attached to a fixed frame by two additional torsion springs that allow for tip-tilt motion along a second axis that is perpendicular to the first axis. Assembly arms lift up the fixed frame, which is attached to hinged sidewalls. The sidewalls are locked into place by dovetails that fit into tapered cuts in the sidewalls. Counter-electrodes are placed below the gimbal ring and the mirror. (Reprinted with permission from IEEE Journal of Lightwave Technology, *Beam-steering micromirrors for large optical cross-connects*, V.A. Aksyuk et al. ©1967 IEEE.)

obtained by etching a thin sacrificial oxide such as the 2 μm thick Oxide1 in the PolyMUMPS process. The frame is lifted out of plane by assembly arms. The assembly arms, the reflective coating for the mirror, and the wire-bond pads use custom metallization.

The torsional springs are formed by folded serpentine polysilicon beams. This spring design does not require fine dimensional control and provides stress relief, making the springs insensitive to small variations in the residual and thermally induced stress in the beams. Both of these factors help to increase the yield and reliability. The mirrors were subjected to 18 billion switching cycles with no detectable change in device performance, indicating that there was no fatigue, creep, or hinge memory effects. The mirrors were also subjected to overvoltages that caused them to tilt sufficiently to touch down on grounded landing pads, but they were able to recover without any degradation or sticking after the voltage was reduced.

To obtain flat mirrors, with a radius of curvature more than 250 mm, the processing conditions for the polysilicon and mirror metallization had to be controlled to minimize residual stresses and stress-gradients. The thickness of the polysilicon mirror membrane was 3.5 μm, a thickness that can be obtained by stacking Poly1 (2 μm) and Poly2 (1.5 μm) in the PolyMUMPS process.

To lift the reflector out of plane after it had been released, a special high-stress metallization layer was deposited on top of 1.5 μm thick, 100 μm wide polysilicon assembly arms. The composition of this high-stress layer (tensile) causes the arms to curl upward after release, lifting the mirror frame and locking it into place. When the assembly arms are released, they curl up into arcs with a radius of curvature R that is dependent on the thin-film stress. The height at a distance x from the anchor of the arm is given by

$$z_{up}(x) = \frac{x^2}{2R}. \quad (4.22)$$

If the end of the arm is acted on by a force F, the arm will be deflected downward by a distance given by

$$z_{down}(x) = -\frac{Fx^2(3L-x)}{6EI}. \quad (4.23)$$

The total height is given by the sum of the two deflections:

$$z(x) = z_{up}(x) + z_{down}(x) = \frac{x^2}{2R} - \frac{Fx^2(3L-x)}{6EI} \quad (4.24)$$

$$z(L) = \frac{L^2}{2R} - \frac{FL^3}{3EI}. \quad (4.25)$$

If the assembly arm lifts the mirror to a height h, it will apply a holding force F given by

$$F = \frac{3EI}{L^3}\left(\frac{L^2}{2R} - h\right). \quad (4.26)$$

A two-axis tip-tilt mirror was designed and fabricated in the PolyMUMPS process as a student design project [12]. Here the mirror was not lifted out of plane as in the lambda router switch, so the maximum tilt angle was determined by the radius of the mirror, 100 μm, set by the dimensions of the optical fiber core that the switch was designed for, and the maximum sacrificial gap of 2.75 μm obtained

4.3 Dual-axis torsional mirror 85

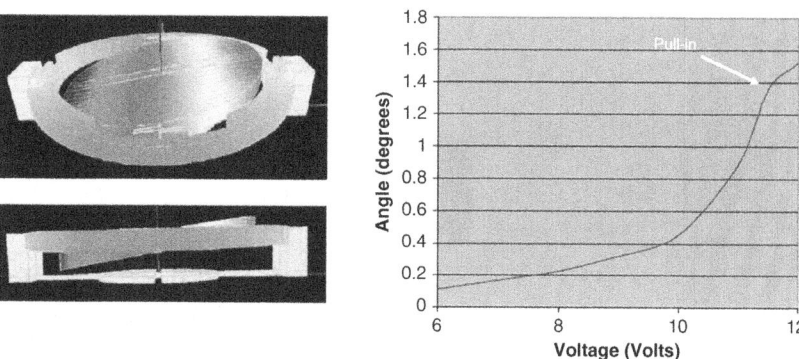

Figure 4.10 Tilt of the mirror about one axis. The deflection of the mirror is shown in the solid model on the left-hand side of the figure and the tilt angle as a function of the applied voltage is shown on the right-hand side of the figure. The pull-in voltage is at approximately 11.5 V, as shown by the arrow in the figure. (Image courtesy of Kuan-Fu Chen and Tung-Chien Chen, EE215 MEMS Design final project report, Spring 2007.) See color plate section.

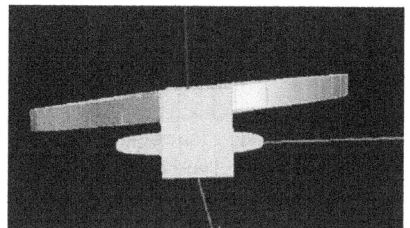 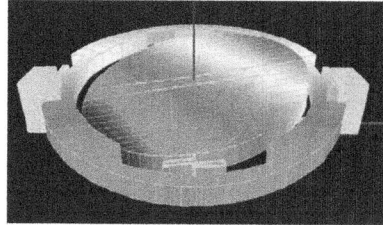

Figure 4.11 (Left) Tip of the mirror by actuation of the ring. (Right) Tip and tilt of the mirror by actuating both the mirror and the ring. (Image courtesy of Kuan-Fu Chen and Tung-Chien Chen, EE215 MEMS Design final project report, Spring 2007.) See color plate section.

by stacking Oxide1 (2 μm) and Oxide2 (0.75 μm). Thus the largest tilt angle was $\theta = \arcsin(2.75/100) = 1.58$ degrees.

The two-axis mirror does not have cylindrical symmetry like the one-axis scanner, so the solution of the tilt angle as a function of the applied voltage is more complex. It can be solved numerically, with finite element analysis, or analytically with some approximations [13]. Here we consider finite element analysis. Tilting of the mirror when it is actuated is shown in Figure 4.10. Here the ring is not actuated, so the ring does not tip. In Figure 4.10 (left), the ring is actuated but the mirror is not, so both the mirror and the ring tip together. In Figure 4.11 (right), both the ring and the mirror are actuated, so the mirror is both tipped and tilted.

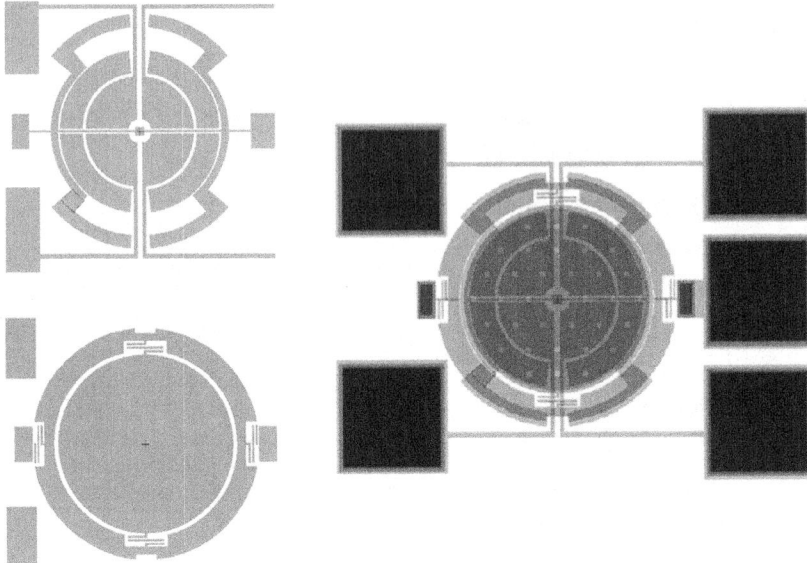

Figure 4.12 Layout for two-axis torsional mirror. In the top left side of the figure the layout for the Poly0 layer is shown. This includes the landing pads and counter-electrodes for the mirror and gimbal ring. The bottom left-hand side of the figure has the layout for the Poly2 layer, which has the mirror and gimbal ring. The right-hand side of the figure shows the full layout including the metallization layer to increase the mirror reflectivity. (Image courtesy of Kuan-Fu Chen and Tung-Chien Chen, EE215 MEMS Design final project report, Spring 2007.)

The layout for the two-axis mirror is shown in Figure 4.12. The mirror is fabricated in Poly2. There are indents at the top and bottom of the outer ring that are used to prevent contact with the Poly0 wires when the outer ring lands. The mirror is connected to the ground potential. Poly0 is used for the electrodes and landing areas. The landing areas prevent the mirror from touching the wires and prevent the mirror from sticking on the substrate. The landing areas are all connected to the ground potential. A picture of the fabricated mirror is shown in Figure 4.13. The issue with print-through of the underlying topography from the counter-electrodes and control wires onto the mirror surface can be seen in this top-down image.

4.4 Fabry-Perot interferometer in the PolyMUMPS process

A Fabry-Perot interferometer is formed by an adjustable optical cavity between two partially reflecting mirrors, as shown schematically in

4.4 Fabry-Perot interferometer in the PolyMUMPS process

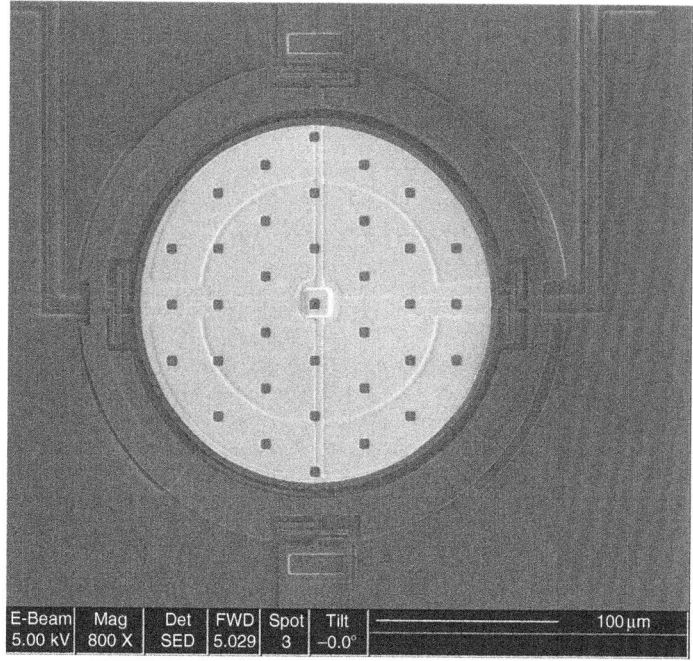

Figure 4.13 Two-axis tip-tilt mirror designed and fabricated as a student project in the PolyMUMPs process. (Image courtesy of Kuan-Fu Chen and Tung-Chien Chen, EE215 MEMS Design final project report, Spring 2007.)

Figure 4.14. Typically the mirrors are fabricated using a Bragg dielectric stack consisting of multilayers of dielectrics with different indices of refraction, n_1 (high) and n_2 (low), stacked alternately with high and low indices of refraction for several cycles. The thickness of each layer is chosen so that a quarter-wavelength of the light that is to be filtered fits in the gap:

$$t_1 = \frac{\lambda}{4n_1} \quad t_2 = \frac{\lambda}{4n_2}. \quad (4.27)$$

The distance between the mirrors can be adjusted with a MEMS actuator to enable a tunable interferometer. When the distance between the mirrors is adjusted to $\lambda/2$, the light passes through the interferometer while light with other wavelengths is blocked. The Fabry-Perot interferometer acts as a narrow bandpass filter for light, with the bandpass is dependent on the "finesse" of the optical cavity.

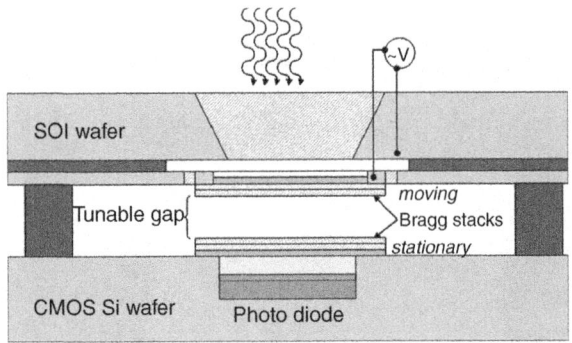

Figure 4.14 Fabry-Perot interferometer fabricated in an SOI technology. Light is incident on the back side of the SOI wafer, which has a through-wafer etch to the device layer on the front side of the wafer. A Bragg dielectric stack composed of transparent thin film layers with alternating high/low indices of refraction are defined in a moveable frame that has been released from the SOI wafer. The frame can be moved by applying a voltage between the substrate and the device layer of the SOI wafer. The wafer with the moveable Bragg stack is bonded to a CMOS wafer that has a stationary stack using a thick gap layer such as polyimide. A photodiode detects light that is transmitted through the Fabry-Perot interferometer.

For light that is incident normally on a Fabry-Perot filter with a mirror separation d and index n in the cavity between the mirrors, the transmission function is given by

$$T = \frac{(1-R)^2}{1+R^2 - 2R\cos(\delta)} \quad \delta = \left(\frac{2\pi}{\lambda}\right) 2nd, \tag{4.28}$$

and the finesse F is given by

$$F = \frac{4R}{(1-R)^2}. \tag{4.29}$$

The larger F is, the narrower the bandpass of the filter.

One of the main challenges with fabricating a Fabry-Perot interferometer is in keeping the Bragg reflectors flat when the thin-film dielectric stack has stresses and stress-gradients, and maintaining the flatness and parallelism of the reflectors when they are actuated to tune the filter. We will consider the design of a Fabry-Perot interferometer in the PolyMUMPS process by Louchis and Hemphill [14]. In this work two designs were considered, a mid-wavelength (MW) IR filter ($\lambda = 1.4\,\mu m$ to $3\,\mu m$) and a short-wavelength IR filter ($\lambda = 3\,\mu m$ to $8\,\mu m$). The mid-wavelength design used stacked Oxide1 ($2\,\mu m$) and Oxide2 ($0.75\,\mu m$)

4.4 Fabry-Perot interferometer in the PolyMUMPS process

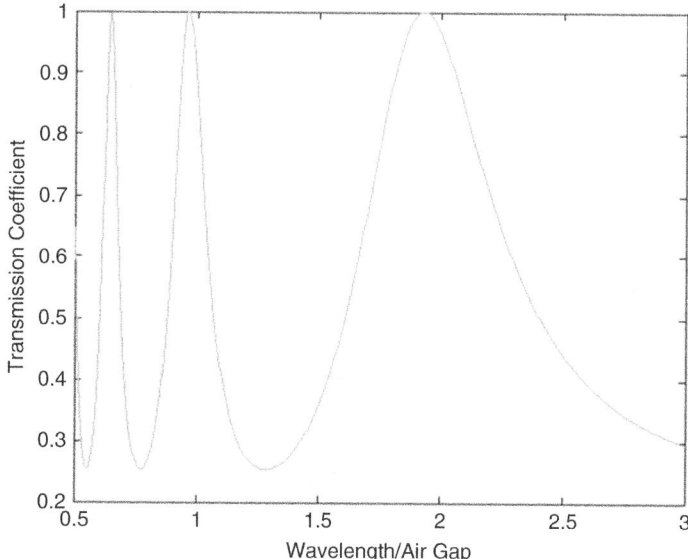

Figure 4.15 Cross-sectional view of the mid-wavelength IR tunable Fabry-Perot interferometer. Layers in descending order from the top are Poly2, Oxide1+ Oxide2, Poly0, nitride, and substrate. (Courtesy of Kevin Louchis and Benjamin Hemphill, Team Ninja Star in the Eye, EE115-Winter 2008, Final Report.)

Figure 4.16 The transmission coefficient of the mid-wavelength IR filter calculated with Matlab as a function of the wavelength normalized to the air gap. (Courtesy of Kevin Louchis and Benjamin Hemphill, Team Ninja Star in the Eye, EE115-Winter 2008, Final Report.)

to get as large a gap as possible in this process. The top reflector was formed in Poly2 (1.5 μm) and the bottom reflector was formed by the stack Poly0 (0.5 μm) on top of nitride (0.6 μm thick) on top of the silicon substrate (525 μm). Undoped silicon is transparent to IR radiation, so it can pass through the substrate so long as the doping is not too high and the layers on the back side of the wafer are not too attenuating. The resulting layer stack is shown in Figure 4.15.

The transmission function T for this filter is plotted in Figure 4.16. As seen in Figure 4.16, the relaxed MW-FPI allows wavelengths around 5.4 μm to be transmitted through the FPI. Figure 4.17 shows the shift in the transmission function as the top mirror is actuated to decrease the

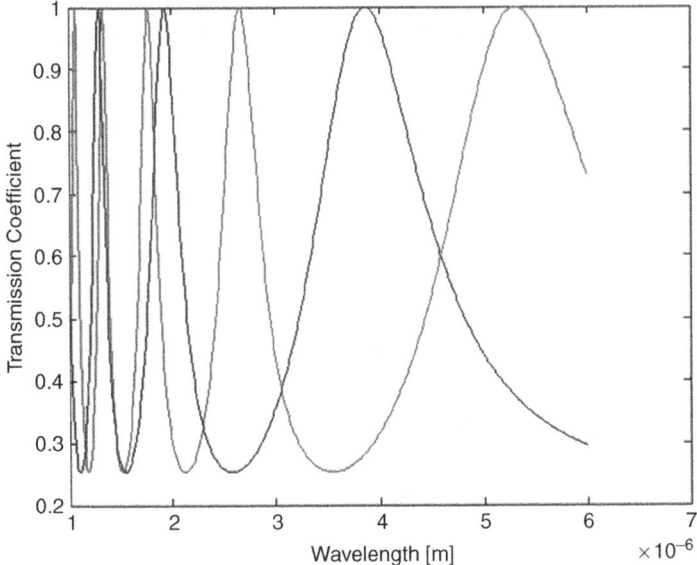

Figure 4.17 The shift in the transmission coefficient of the mid-wavelength IR filter calculated with Matlab as a function of the wavelength. (Courtesy of Kevin Louchis and Benjamin Hemphill, Team Ninja Star in the Eye, EE115-Winter 2008, Final Report.) See color plate section.

Figure 4.18 Short-wavelength interferometer with top Bragg reflector using Oxide2 trapped between Poly1 and Poly2. (Courtesy of Kevin Louchis and Benjamin Hemphill, Team Ninja Star in the Eye, EE115-Winter 2008, Final Report.)

gap. As the Poly2 layer is pulled in by the voltage applied on the Poly0 layer, the maximum transmission coefficient shifts from 5.4 μm to 3.9 μm as the gap narrows.

The second design is for a short-wavelength tunable IR filter. In this design the initial gap was defined by the 2 μm thick sacrificial Oxide1, and the second layer of oxide, Oxide2, which is 0.75 μm thick, was trapped between Poly1 and Poly2, forming a dielectric stack Poly2/Oxide2/Poly1 to increase the reflectivity of the top mirror as shown in the cross section in Figure 4.18.

While not the optimal layer thickness for the Bragg stack, which would have dielectric layer thicknesses that are one-quarter of the wavelength of light that is to be filtered, it makes good use of the available layers in the

4.4 Fabry-Perot interferometer in the PolyMUMPS process

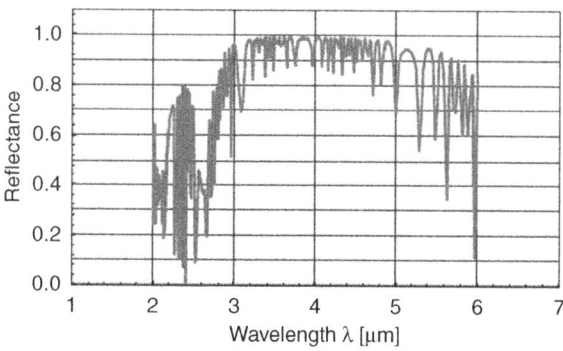

Figure 4.19 InfraTec's experimental data giving the reflectance of a polysilicon–SiO$_2$–polysilicon dielectric stack. (Courtesy of Kevin Louchis and Benjamin Hemphill, Team Ninja Star in the Eye, EE115-Winter 2008, Final Report.)

PolyMUMPs process. Nonetheless, the Poly2 and Oxide2 layers are relatively close to a quarter-wavelength of mid-band light in our range, and the Poly1 layer is at one-half the mid-band wavelength. The trapped oxide also makes the top reflector thicker (2 μm Poly1, 0.75 μm Oxide1, and 1.5 μm Poly2 for a total thickness of 4.25 μm) and thus stiffer to help maintain flatness. The bottom reflector is formed from Poly0, 0.5 μm thick, on top of nitride, 0.6 μm thick, on top of the substrate. Lightly doped silicon is transparent to wavelengths longer than 1 μm, so the IR light is able to pass through the lightly doped substrate. Figure 4.19 shows the reflectivity of a poly/oxide/poly stack used by InfraTec in their tunable IR spectrometer [15]. They were able to adjust the layer thickness in their process to optimize the reflectivity.

These data were used to update the reflectance for the transfer function of the short-wavelength filter starting from a relaxed state with a 2 μm gap that is decreased to a 1.5 μm gap by electrostatic forces between the Poly0 layer and the dielectric stack The new transmission function is plotted in Figure 4.20. It can be seen that the improved dielectric stack increases the finesse and sharpens the transmission peaks.

Solid models of the short- and long-wavelength interferometer designs are shown in Figure 4.21. Both designs use folded springs to decrease their stiffness, enabling actuation using CMOS-compatible voltages.

The top mirror support is four springs acting in series. Each individual spring is comprised of a set of three fixed-guides beams connected end-to-end and acting in parallel.

92   Optical MEMS

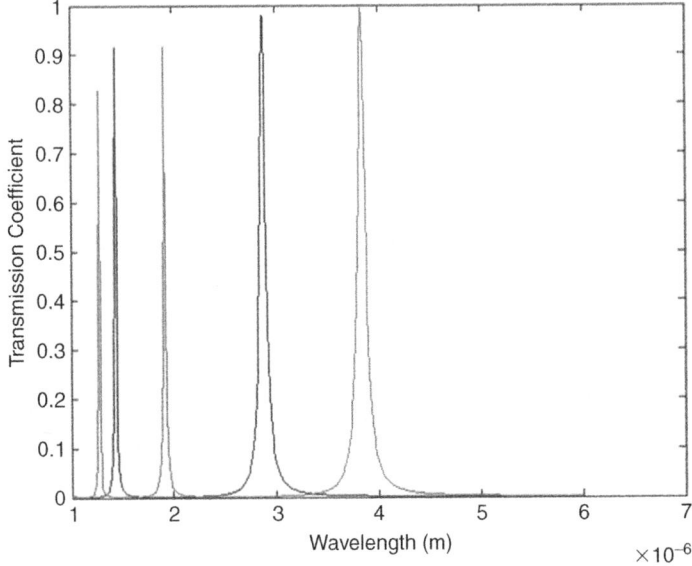

Figure 4.20 Transmission function calculated for the short-wavelength filter using the reflectance data from Figure 4.19. The red plot shows the transmission function for the short wavelength when the air gap is in the relaxed position, and the blue plot shows the transmission function after the air gap has been reduced by 0.5 μm. (Courtesy of Kevin Louchis and Benjamin Hemphill, Team Ninja Star in the Eye, EE115-Winter 2008, Final Report.) See color plate section.

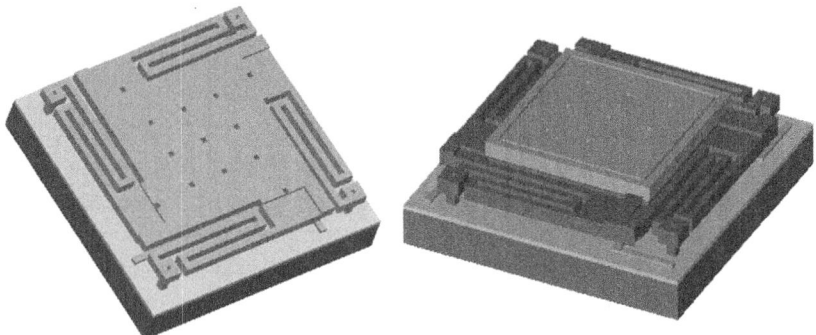

Figure 4.21 Solid models of the long-wavelength tunable Fabry-Perot interferometer (left) and the short-wavelength interferometer (right). The short-wavelength interferometer has a Bragg stack consisting of Oxide2 trapped between Poly1 and Poly2 for the top reflector. (Courtesy of Kevin Louchis and Benjamin Hemphill, Team Ninja Star in the Eye, EE115-Winter 2008, Final Report.)

4.5 Obtaining flatness in optical MEMS devices

An important consideration for optical MEMS devices is how flat the optical surface can be; a mirror surface with a radius of curvature R functions as a lens with a focal length $f = R/2$ rather than as a mirror. In addition to curvature, the roughness of a surface will impact the optical performance of a mirror, and often specifications will call for a root mean square surface roughness of less than $\lambda/20$. For light with a wavelength of 0.5 µm, this specification would be for a root mean square roughness of less than 25 nm. In some demanding optical applications in astronomy, the specifications call for less than 1 nm surface roughness.

An uncoated optical surface can be deformed from stress and stress-gradients. For example, a membrane with sufficient compressive residual stress can buckle if it is constrained by the boundary conditions. For a constrained surface a slightly tensile stress is preferred, where the membrane is pulled flat by the boundary conditions like the skin on a drum. If there is a stress-gradient, the optical membrane will deform when it is released [16], [17]. The stress-gradient can be due to the thin-film growth process itself or generated by the deposition of a thin film on one side of the mirror, such as a reflective metal layer that has stress, stress-gradients, or a different coefficient of thermal expansion (CTE). The CTE mismatch will create a stress gradient on cooling if the metallization is deposited at an elevated temperature. In addition, the released mirror will deform like a bimorph if it is heated or cooled. In general, a thick mirror and a thin metallization layer are preferred to minimize deformations due to stress and stress-gradients. The 10–25 µm thick device layer of a silicon on insulator (SOI) wafer can be useful for fabricating flat optical mirrors [18], [19]. Thus the SOIMUMP process is ideal for prototyping optical MEMS devices with a flat surface. The 2 µm thick Poly1 or 1.5 µm thick Poly2 layer in the MUMPS process with a gold metallization layer (500 nm Au/ 20 nm Cr) to increase reflectivity can become significantly distorted on release. The metallization layer in the PolyMUMPS process can have 50–100 MPa of residual stress, which can significantly distort the thin poly layers when they are released. An example of a released metallized Poly1 surface in the PolyMUMPS process that has been distorted is shown in Figure 4.22.

In addition to deformations due to stress and stress-gradients, the surface roughness of the thin films used to fabricate optical MEMS components can also have an impact, as deposited polysilicon can have significant roughness. The root mean square surface roughness of polysilicon in the MUMPS process has been measured and found to be approximately 12 nm.

Figure 4.22 Released metallized mirror in the PolyMUMPS process. The mirror is distorted due to the tensile stress of the metallization.

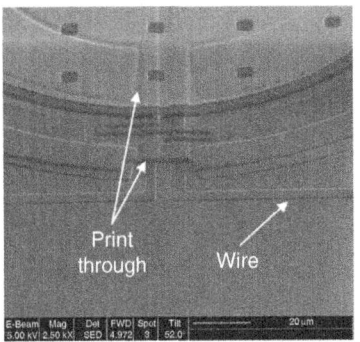
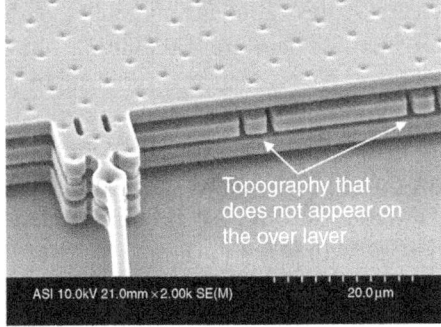

Figure 4.23 Print-through of the wires and counter-electrodes from underlying layers in the PolyMUMPS process (left). Mirror fabricated by MEMX in the SUMMiT process that uses chemical mechanical polishing to planarize the topography that develops from conformal coatings (right).

Additional topography can be generated from conformal coatings, where the topography associated with structures in underlayers generates surface topography in overlayers. An example of the print-through of underlying topography in the PolyMUMPS process is shown in Figure 4.23 (left).

4.5 Obtaining flatness in optical MEMS devices

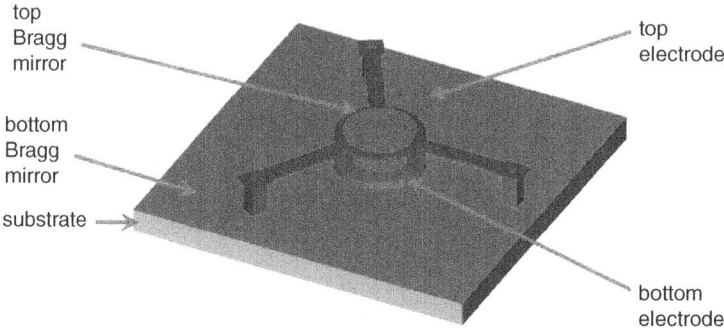

Figure 4.24 Tunable Fabry-Perot optical filter. (Figure courtesy of Ms. Mary Ann Maher, SoftMEMS.)

The SUMMiT process uses chemical mechanical polishing to planarize the topography generated by features in the lower layers, as can be seen in Figure 4.23 (right).

Problems

(1) Lay out a tunable Fabry-Perot optical filter in a multiproject wafer process as shown in Figure 4.24. The top Bragg mirror should be composed of a stack of dielectric films and should be flat to within a fraction of a wavelength of light that will be filtered.

(2) Design a 10×10 MEMS deformable mirror array for adaptive optics in the visible wavelength band. The design should provide at least 0.75 μm of mechanical stroke and have mirrors that are flat to within $\lambda/20$. The operating voltage should be less than 275 V to avoid breakdown. Before starting your design, do a literature search to see what previous designs have been fabricated.
  a. Design a segmented mirror.
  b. Design a mirror with a continuous face sheet.

(3) Design a MEMS-based scanner for a cell phone projection display. Before starting your design, do a literature search to see what previous designs have been fabricated.

(4) Design a MEMS display for a cell phone. Before starting your design, do a literature search to see what previous designs have been fabricated.

REFERENCES

1. K.E. Petersen, *Silicon as a mechanical material*, Proc. IEEE 70, pp. 420–457 (1982).
2. K.E. Petersen, *Dynamic micromechanics on silicon: techniques and devices*, IEEE Trans. Electron Devices ED-25(10), pp. 1241–1250 (1978).
3. G.T.A. Kovacs, *Micromachined Transducers Sourcebook*, McGraw-Hill (1998).
4. F. Pan, J.A. Kubby, E. Peeters, J. Chen, O. Vitomirov, D. Taylor, and S. Mukherjee, *Design, modeling and verification of MEMS silicon torsion mirror*, Proc. SPIE 3226, p. 114 (1997).
5. F. Pan, J. Kubby, E. Peeters, A.T. Tran, and S. Mukherjee, *Squeeze film damping effect on the dynamic response of a MEMS torsion mirror*, J. Micromechanics Microengineering 8, pp. 200–208 (1998).
6. F. Pan, J. Kubby, E. Peeters, J.K. Chen, O. Vitomirov, D. Taylor, and S. Mukherjee, *Design, modeling and verification of a MEMS silicon torsion mirror for applications in xerography*, International Instrumentation Symposium, 44th, Reno, NV, 3–7 May, pp. 66–75 (1998).
7. Stephen D. Senturia, *Microsystem Design*, Kluwer Academic Publishers, Boston/Dordrecht/London (2001). ISBN 0-7923-7246-8.
8. V.A. Aksyuk, F. Pardo, C.A. Boll, S. Arney, C.R. Giles, and D.J. Bishop, *Lucent Microstar micromirror array technology for large optical crossconnects*, Proc. SPIE 4178, pp. 320–324, MOEMS and Miniaturized Systems, M.E. Motamedi and R. Goering, Eds., pp. 320–324 (2000).
9. R. Ryf, D.T. Neilson, and C.R. Giles, *Scalable micro mechanical optical crossconnects*, Proc. SPIE 4455, Micro- and nano-optics for optical interconnection and information processing, M.R. Taghizadeh, H. Thienpont, and G.E. Jabbour, Eds., December, pp. 51–58 (2001).
10. V.A. Aksyuk, F. Pardo, D. Carr, et al., *Beam-steering micromirrors for large optical cross-connects*, J. Lightwave Technology 21(3), pp. 634–642 (2003).
11. M. Mizukami, J. Yamaguchi, N. Nemoto, et al., *128×128 3D-MEMS optical switch module with simultaneous optical paths connection for optical cross-connect systems*, IEEE Conf. Proc. Photonics in Switching, 2009, PS '09, Pisa, Italy, Sept., pp. 1–2 (2009).
12. K.-F. Chen, and T.-C. Chen, *Optical Cross-connect Switch (OXC)*, EE215 MEMS Design, UC Santa Cruz, Spring 2007.
13. Z. Xiao, W. Peng, X.-T. Wu, and K.R. Farmer, *Pull-in study for round double-gimbaled electrostatic torsion actuators*, J. Micromechanics Microengineering 12, 77–81 (2002).
14. K. Louchis, and B. Hemphill, *Team Ninja Star in the Eye*, EE115, Final Report, Winter (2008).
15. N. Neumann, M. Heinze, S. Kurth, and K. Hiller, *A Tunable Fabry-Perot-Interferometer for 3–4.5 µm Wavelength with Bulk Micromachined Reflector Carrier*. InfraTec.
16. V.A. Aksyuk, F. Pardo, and D.J. Bishop, *Stress-induced curvature engineering in surface micromachined devices*, Proc. SPIE 3680, Design, Test, and Microfabrication of MEMS and MOEMS, B. Courtois, S.B. Crary, W. Ehrfeld, et al., Eds., March, pp. 984–993 (1999).

17 T.G. Bifano, H.T. Johnson, P. Bierden, and R.K. Mali, *Elimination of stress-induced curvature in thin-film structures*, J. Microelectromechanical Systems 11(5), pp. 592–597 (2002).
18 J.A. Kubby on behalf of the MOEMS Manufacturing Consortium, *Hybrid integration of light emitters and detectors with SOI-based micro-opto-electromechanical (MOEMS) systems,* Proc. SPIE 4293, Silicon-Based and Hybrid Optoelectronics III, D.J. Robbins, J.A. Trezza, and G.E. Jabbour, Eds., pp. 32–45 (2001).
19 J.A. Kubby, *Hybrid silicon-on-insulator micromachining for critical MEMS components*, Solid State Technology 47(9), September (2004).

5
Thermal MEMS

Thermal MEMS make use of heat transfer for their operation. Some of the modes of heat transfer that are used include conduction through a solid, convection through a gas or a liquid, and optical radiation at high temperatures. Heat flow through a long, thin beam is driven by a temperature gradient ΔT, where heat Q flows from the hot end of the beam toward the cold end, as shown schematically in Figure 5.1.

The heat transfer rate, $\partial Q/\partial t$, depends on the cross-sectional area A of the beam and its thermal conductivity κ according to Fourier's Law:

$$\frac{\partial Q}{\partial t} = -\kappa A \frac{\partial T}{\partial x}. \qquad (5.1)$$

The thermal conductivity has units of W/m·K. The minus sign is because heat flows downhill, from hot to cold. An analogy can be made to current flow I through a resistor R. Here the heat flow, $\partial Q/\partial t$, corresponds to the current flow I and the temperature drop ΔT corresponds to the voltage drop ΔV. In this electrical analogy the thermal conductivity κ corresponds to the electrical conductivity, σ. For a long thin beam of length L with a temperature drop ΔT across it, the analogy can be written as

$$\frac{\Delta Q}{\Delta t} = -\left(\frac{\kappa A}{L}\right)\Delta T \quad \text{(thermal)} \Leftrightarrow I = \frac{\Delta V}{R} = \left(\frac{A}{\rho L}\right)\Delta V = \sigma \Delta V \quad \text{(electrical)}, \qquad (5.2)$$

where ρ is the electrical resistivity. In this analogy, $L/\kappa A$ can be thought of as a thermal resistance, $\Delta Q/\Delta t$ as a thermal current, and ΔT as a thermal potential. A lumped parameter model listing the analogies between the thermal and electrical domains is provided in Table 5.1.

Heat transfer from convection occurs in a medium due to molecular motion in a fluid or gas. The flow can be natural, from changes in density

Table 5.1 *Lumped parameter model for thermal MEMS*

	Lumped parameter model	
	Thermal	Electrical
Driving force	ΔT (temperature)	ΔV (voltage)
Coordinate	Heat Q	Charge q
Flow	Heat flux $J = \partial Q/\partial t$	Electrical current $I = \partial q/\partial t$
Capacitance	Heat capacity C	Electrical capacity C
Conductivity	Thermal conductivity κ	Electrical conductivity σ

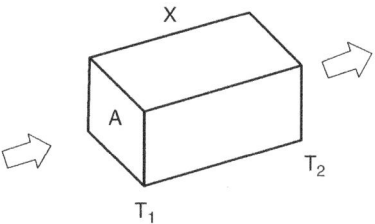

Figure 5.1 Beam element of length x and cross-sectional area A. One end of the beam is at T_1 and the other end is at T_2, where $T_1 > T_2$. Heat flows from the hot side to the cold side. (Reprinted from University of Plymouth BEng Stage 2 course notes with permission from Dr. Murray A. Bell.)

with changes in temperature, or forced, from an actuator like a pump. Newton's law of cooling describes convective heat transport between the surfaces of a hot body to a fluid, as shown in Figure 5.2.

The heat transfer is proportional to the area of the surface–fluid interface and the temperature difference between the surface, $T_{surface}$, and the fluid, T_{fluid}. The constant of proportionality is called the surface heat transfer coefficient h:

$$\frac{\partial Q}{\partial t} = hA\left(T_{surface} - T_{fluid}\right). \qquad (5.3)$$

Finally, another important heat transfer mechanism is through radiation. Here the heat transfer is determined by the Stephan-Boltzmann Law that relates the heat flow to the temperature of the radiating body:

$$\frac{\partial Q}{\partial t} = \varepsilon\sigma T^4, \qquad (5.4)$$

where ε is the emissivity of the surface of the radiator and is related to the reflectivity R and the absorptivity A by

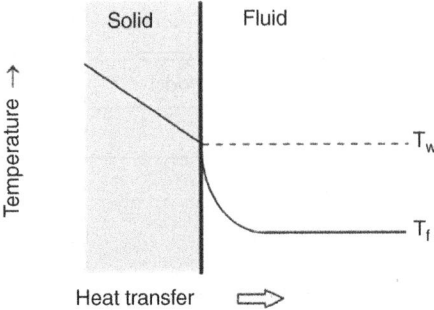

Figure 5.2 Heat transfer from a solid to a fluid by convection. At the boundary between the solid and the fluid, the wall temperature is T_W. Far from the wall the temperature of the fluid is T_f. As the fluid heats up its density changes, giving rise to a buoyant force that causes fluid motion. (Reprinted from University of Plymouth BEng Stage 2 course notes with permission from Dr. Murray A. Bell.)

$$R = 1 - \varepsilon = 1 - A. \tag{5.5}$$

A perfect "black-body" that absorbs all of the radiation that is incident on it and does not reflect any light has $\varepsilon = 1$. In general a "grey-body" has $0 < \varepsilon < 1$, white paint has $\varepsilon = 0.95$, and polished steel has $\varepsilon = 0.07$. The Stephen-Boltzmann constant σ is given by

$$\sigma = 5.67 \times 10^{-8} \frac{W}{m^2 K}. \tag{5.6}$$

Since the heat flow is proportional to the fourth power of the temperature, the heat flow due to radiation is most important at high temperatures.

In addition to heat transfer, there can be sources of heat within a MEMS device. One of the more common heat sources is Joule heating, where the power dissipated by a resistance R carrying a current I is given by

$$\frac{\partial Q}{\partial t} = I^2 R. \tag{5.7}$$

In general, all of these heat transfer mechanisms can occur. The power balance equation is given by

$$\frac{\partial \tilde{Q}}{\partial t} = \tilde{P}_{Sources} - \nabla \cdot J_Q, \tag{5.8}$$

where $\tilde{Q}$ is the heat energy in the material, $\tilde{P}_{Sources}$ is the heat generated in the material, and J_Q is the heat flux out of the material. The tilde ($\sim$) indicates that the heat and the heat power generated have been normalized to a unit volume of the material in this equation. Fourier's Law is given by

$$J_Q = -\kappa \nabla T, \tag{5.9}$$

where κ is the thermal conductivity. Taking the gradient of the heat flux gives

$$\nabla \cdot J_Q = -\nabla \cdot \kappa \nabla T = -\kappa \nabla^2 T, \tag{5.10}$$

and the power balance equation becomes

$$\frac{\partial \tilde{Q}}{\partial t} = \tilde{P}_{Sources} + \kappa \nabla^2 T \tag{5.11}$$

$$\frac{\partial \tilde{Q}}{\partial t} - \kappa \nabla^2 T = \tilde{P}_{Sources}. \tag{5.12}$$

If we make the substitution for the heat capacity, then

$$\tilde{C} = \frac{\partial \tilde{Q}}{\partial T} \tag{5.13}$$

$$\tilde{C} = \frac{\partial \tilde{Q}}{\partial T} \rightarrow \partial \tilde{Q} = \tilde{C} \partial T \rightarrow \frac{\partial \tilde{Q}}{\partial t} = \tilde{C} \frac{\partial T}{\partial t}. \tag{5.14}$$

If this is substituted into the power balance equation,

$$\tilde{C} \frac{\partial T}{\partial t} - \kappa \nabla^2 T = \tilde{P}_{Sources} \tag{5.15}$$

$$\frac{\partial T}{\partial t} = \frac{\kappa}{\tilde{C}} \nabla^2 T + \frac{\tilde{P}_{Sources}}{\tilde{C}}. \tag{5.16}$$

In thermal equilibrium where $\partial Q/\partial t = 0$, the heat leaving a system, Q_{out}, is equal to the heat entering the system, Q_{in}, and the heat balance equation is given by

$$Q_{Joule\,heating} + Q_{radiation\,absorbed} = Q_{radiation\,out} + Q_{conduction} + Q_{convection}. \tag{5.17}$$

5.1 Thermal actuator

Thermal actuation makes use of the thermal expansion of solids as they are heated, or the differences in the rates of thermal expansion between different materials as in a bimorph actuator. Relative to electrostatic

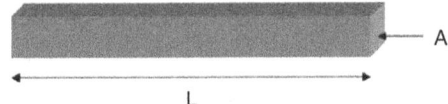

Figure 5.3 A bar of length L and cross-sectional area A is heated by passing a current through it, causing it to expand by a length ΔL if it is unconstrained. If the bar is constrained, the thermal expansion will give rise to a stress σ and a force $F = \sigma A$.

actuation, thermal actuation requires lower voltages, on the order of 10–25 V, and can generate larger forces, on the order of mN. Thermal actuators typically use Joule heating, where a current is passed through a resistor, requiring relatively high currents (mA) and the dissipation of continuous power (mW) to maintain their positions. Electrostatic actuators require relatively high voltages, on the order of 10–100 V, but they do not require high continuous current levels to maintain their positions and only dissipate power when the electrostatic actuator is charged or discharged. They generate smaller forces relative to thermal actuators, on the order of µN. They are used where lower power and lower forces are required.

In a resistive thermal actuator, Joule heating causes an actuator material to heat up by an amount ΔT. The heating leads to thermal expansion, which can be used to create a deflection for an actuator. If we consider the thermal expansion of a bar of length L, as shown in Figure 5.3, as the bar is heated by an amount ΔT it will expand by an amount ΔL, giving rise to a strain ε, where

$$\varepsilon = \frac{\Delta L}{L} = \gamma \, \Delta T, \qquad (5.18)$$

where γ is the coefficient of thermal expansion (CTE) of the material. If the bar is constrained, the thermal expansion will give rise to a stress σ, where

$$\sigma = E\varepsilon, \qquad (5.19)$$

where E is the Young's modulus of the material. The stress σ will give rise to a force F, where

$$F = A\sigma, \qquad (5.20)$$

where the bar has a cross-sectional area A. As an example we can calculate the strain, stress, force, and length change of a silicon beam that is 100 µm long, 2 µm wide, and 2 µm thick. The cross-sectional area would then be 4×10^{-12} m². If the beam is heated by 100 K, then the strain ε would be

$$\varepsilon = \gamma \Delta T = (2.3 \times 10^{-6}/\text{K})(100\,\text{K}) = 2.3 \times 10^{-4}. \tag{5.21}$$

If the beam is constrained, the stress σ would be

$$\sigma = E\varepsilon = (160\,\text{GPa})(2.3 \times 10^{-4}) = 36.8\,\text{MPa}. \tag{5.22}$$

The force F on the constrained beam would be

$$F = A\sigma = (4 \times 10^{-12}\,\text{m}^2)(36.8 \times 10^6\,\text{Pa}) = 1.47 \times 10^{-4}\,\text{N}. \tag{5.23}$$

If the beam is not constrained, its length would change by ΔL:

$$\Delta L = \varepsilon L = (2.3 \times 10^{-4})(100\,\mu\text{m}) = 2.3 \times 10^{-2}\,\mu\text{m} = 0.23\,\text{nm}. \tag{5.24}$$

The length does not change significantly, so a useful actuator will require a means to trade off the large force for the small displacement. One approach, the heatuator, converts the small linear change in length to a larger rotational motion of a beam.

5.2 Heatuator

A diagram of a heatuator is shown in Figure 5.4. The heatuator consists of two arms: a narrow hot arm and a wide cold arm that are attached at their far ends. The cold arm also has a short flexure that allows it to rotate around an anchor point near the bond pads. By passing a current through the heatuator, Joule heating causes the narrow hot arm, which

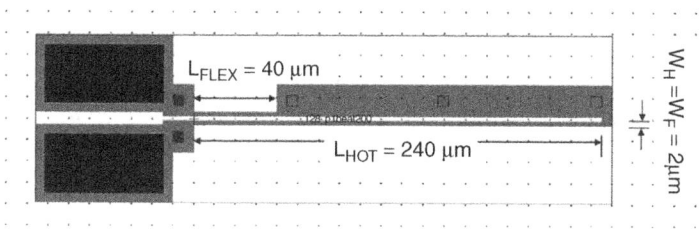

Figure 5.4 A heatuator comprised of a narrow hot arm and a wider cold arm. Current passing through the actuator causes the hot arm to expand more than the cold arm, causing the heatuator to deflect in a circular arc about its attachment point to the substrate at the bond pads.

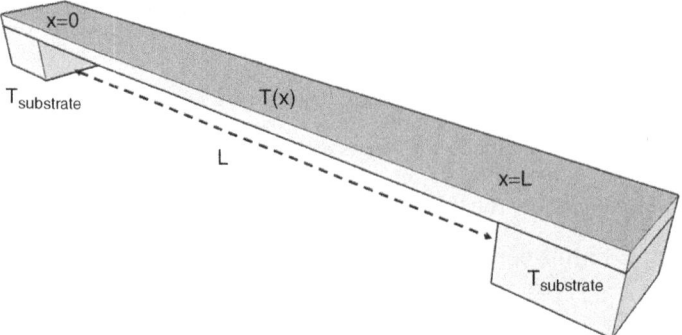

Figure 5.5 Diagram for the temperature distribution $T(x)$ of a clamped-clamped beam.

has a higher resistance, to be heated more than the wider cold arm, which has a lower resistance.

As the hot arm expands it causes the heatuator to swing in an arc about its attachment point to the substrate. The arc length is determined by the difference in length of the hot and cold arms and the overall length of the heatuator. While the thermal expansion gives rise only to a small increase in the difference between their lengths, the overall length of the heatuator can be long, giving rise to several microns of deflection about the center of rotation.

To estimate the difference in length between the hot and cold arms of the heatuator, we will make the approximation that the cold arm stays near the substrate temperature and the hot arm and flexure heat equally along their common length, so that there is no relative change in length between them. We can then find the temperature distribution along the length of a clamped-clamped beam as a first approximation for the temperature rise of the hot arm.

A diagram of the clamped-clamped beam of length L is shown in Figure 5.5. The temperature of the beam at $x=0$ and $x=L$ is $T_{substrate}$. As a current is passed through the beam it heats up and develops a temperature distribution $T(x)$ along its length. The steady state temperature distribution, with $\partial T/\partial t = 0$, can be found from integrating the equation:

$$\frac{\partial T}{\partial t} = 0 = \frac{\kappa}{\tilde{C}}\nabla^2 T + \frac{\tilde{P}_{Sources}}{\tilde{C}} \rightarrow \kappa\nabla^2 T = \tilde{P}_{Sources}. \quad (5.25)$$

In one dimension, $\nabla^2 T = d^2T/dx^2$:

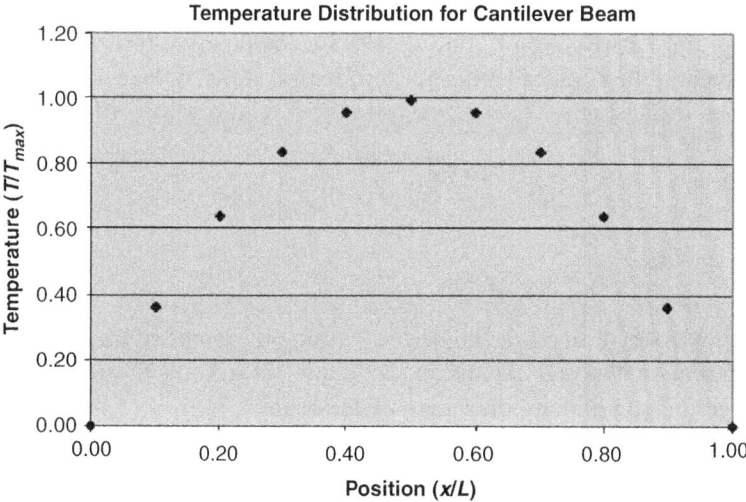

Figure 5.6 Plot of the temperature distribution for the clamped-clamped beam.

$$\frac{d^2T}{dx^2} = -\frac{\tilde{P}_{Sources}}{\kappa}. \tag{5.26}$$

Integrating once,

$$\int \frac{d}{dx}\left(\frac{dT}{dx}\right)dx = \frac{dT}{dx} = -\frac{\tilde{P}_{Sources}}{\kappa}\int dx = -\frac{\tilde{P}_{Sources}}{\kappa}x + b. \tag{5.27}$$

Integrating twice,

$$\int \left(\frac{dT}{dx}\right)dx = T(x) = -\frac{\tilde{P}_{Sources}}{2\kappa}x^2 + bx + c. \tag{5.28}$$

Applying the boundary condition $T(x=0) = T_{substrate}$ gives $c = T_{substrate}$. Applying the boundary condition $T(x=L) = T_{substrate}$ gives

$$T(x=L) = T_{substrate} = -\frac{\tilde{P}_{Sources}}{2\kappa}L^2 + bL + T_{substrate} \rightarrow b = \frac{\tilde{P}_{Sources}}{2\kappa}L. \tag{5.29}$$

Solving for $T(x)$,

$$T(x) = -\frac{\tilde{P}_{Sources}}{2\kappa}x^2 + \frac{\tilde{P}_{Sources}}{2\kappa}Lx + T_{substrate}. \tag{5.30}$$

This temperature distribution is shown in Figure 5.6 for $T_{substrate} = 0$. As expected, the center of the beam is the hottest point because it is furthest from the thermal connections to the substrate.

For approximating the displacement of the heatuator, we can use this temperature distribution to find an average temperature for the hot arm, and then multiply the average hot arm temperature by the coefficient of thermal expansion γ:

$$\Delta T_{average} = \frac{1}{L} \int_{L=0}^{L} (T(x) - T_{substrate}) dx \tag{5.31}$$

$$\Delta L \approx \gamma L \, \Delta T_{average}. \tag{5.32}$$

This approximation can be improved by using the temperature-dependent coefficient of thermal expansion, $\gamma(T)$, and integrating the incremental changes in length along the length of the beam.

The temperature distribution for the heatuator can also be solved using a coupled thermomechanical analysis. *Thermal electrical/thermal stress* involves two types of analyses. The first step is to solve the coupled thermal electrical equations. The coupling arises from internal heat generation, which is a function of the electrical current density (Joule heating). The second step is to perform a thermal stress analysis. The thermal loads are those calculated from the previous step. This assumes that there is no coupling between the thermal-electrical phenomenon of the device and the mechanical deformation. The *heat transfer/thermal stress analysis* solves the three-dimensional heat transfer equation by first using a finite element method to derive the temperature distribution in the structure and then by using this distributed temperature to perform an elastic thermal stress mechanical analysis. Other than the material properties of a structure, the applied temperature gradient, the thermal expansion coefficient, and the applied boundary conditions are the major factors affecting device performance. This analysis produces results for the temperature distribution, the thermal stress distribution, the displacement distribution, and the deformation of the structure.

In the analysis of the heatuator, its resistance is temperature dependent and is given by

$$\rho(T) = \rho_0 (1 + 8.3 \times 10^{-4} \, T + 5 \times 10^{-7} \, T^2),$$
$$\text{where } \rho_0 = 2 \times 10^{-3} \, \Omega/\text{cm}. \tag{5.33}$$

The boundary conditions are applied to the model, mechanically fixing the boundaries where the heatuator is attached to the substrate and specifying the heat loads. In this case the heat loads are specified as convective heat transfers on all of the faces of the heatuator. A voltage

5.2 Heatuator

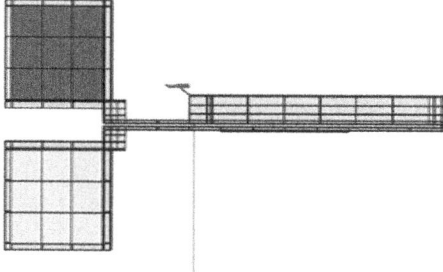

Figure 5.7 Model for the heatuator. The voltage is applied between the two bond pads, causing a current to flow through the hot and cold arms. The hot arm has a higher resistance because it has a smaller cross-sectional area than the cold arm, causing it to heat to a higher temperature. It expands more than the cold arm, causing the heatuator to deflect in an arc about its connection point to the substrate. (Reprinted with permission from IntelliSuite v8.6 (2010).)

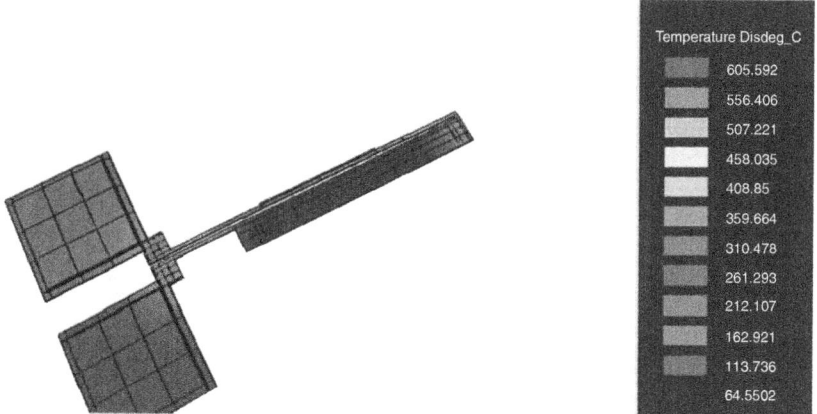

Figure 5.8 Temperature distribution for the heatuator. The hot arm heats up to more than 600°C. (Reprinted with permission from IntelliSuite v8.6 (2010).) See color plate section.

load is then applied between the two bond pads, inducing a current to flow through the heatuator and causing it to deform as the hot arm expands more than the cold arm (Figures 5.7–5.9).

Another type of thermal actuator is the bent-beam thermal actuator shown in Figure 5.10. This actuator is similar to the bent-beam strain sensor [1], but here the strain ε is caused by heating of the beams rather

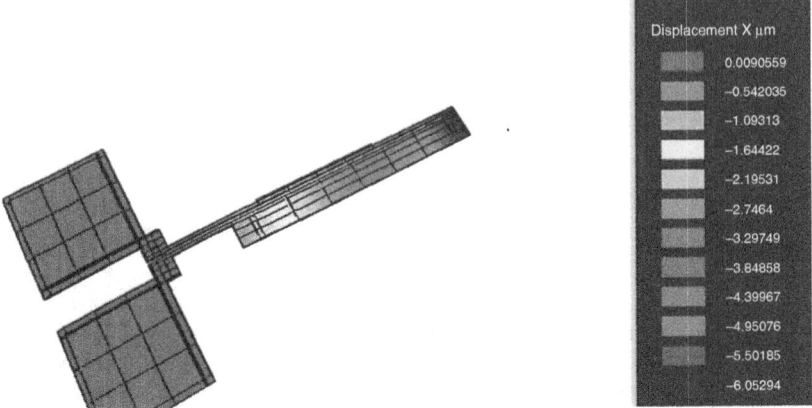

Figure 5.9 Displacement of the heatuator. The tip of the heatuator deflects by more than 6 μm. (Reprinted with permission from IntelliSuite v8.6 (2010).) See color plate section.

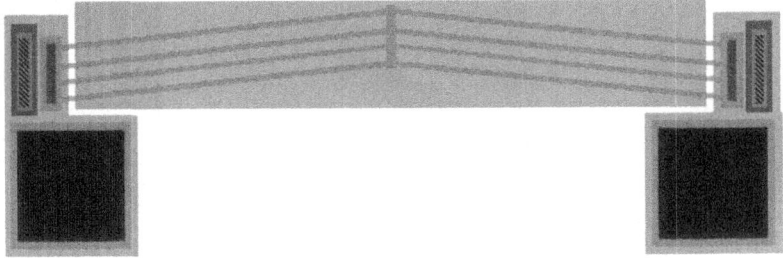

Figure 5.10 Bent-beam thermal actuator. Multiple clamped-clamped beams are ganged together to increase the overall force. As the beams are heated, they displace in the direction of the offset.

than the residual stresses of the beams. This actuator uses clamped-clamped beams that are slightly offset to bias the direction the beams go in when heated. In contrast to the heatuator, which swings in an arc as it displaces, the bent-beam thermal actuator has a linear displacement. In this actuator, multiple clamped-clamped beams are ganged together to increase the total force of the actuator.

The displacement of the beam can be calculated by approximating a right triangle, as shown in Figure 5.11 and solving for the displacement Δx using the Pythagorean theorem:

5.3 Thermal bimorph

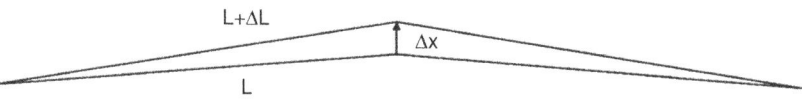

Figure 5.11 Analysis of the bent-beam thermal actuator. When a beam of length L is heated, it expands by a length ΔL, causing the actuator to displace a distance Δx.

$$\Delta x^2 + L^2 = (L + \Delta L)^2$$
$$\Delta x^2 = 2L\Delta L + \Delta L^2$$
$$\approx 2L\Delta L$$
$$\Delta x \approx \sqrt{2L\Delta L}. \tag{5.34}$$

5.3 Thermal bimorph

A thermal bimorph actuator is comprised of two materials with different coefficients of thermal expansion that are bonded together, as shown in Figure 5.12. As the materials heat up they expand at different rates, causing the actuator to deflect toward the side that expands the least.

To solve for the temperature distribution $T(x)$ along the length of the beam, we set the anchored end of the beam at $x=0$ to the temperature of the substrate, T_{sub}. The other end of the beam that is released has the longest thermal path to the substrate, so it will get the hottest. Since the temperature is maximum at $x=L$, the derivative of the temperature distribution will be zero. In equilibrium, $\partial T/\partial t = 0$, so that equation (5.16) becomes

$$\frac{\partial T}{\partial t} = 0 = \frac{\kappa}{\tilde{C}}\nabla^2 T + \frac{\tilde{P}_{Sources}}{\tilde{C}} \tag{5.35}$$

$$-\kappa\nabla^2 T = \tilde{P}_{Sources}. \tag{5.36}$$

In one dimension this becomes

$$\frac{d^2 T}{dx^2} = -\frac{\tilde{P}_{Sources}}{\kappa}. \tag{5.37}$$

Integrating once,

$$\int \frac{d}{dx}\left(\frac{dT}{dx}\right)dx = \frac{dT}{dx} = -\frac{\tilde{P}_{Sources}}{\kappa}\int dx = -\frac{\tilde{P}_{Sources}}{\kappa}x + b. \tag{5.38}$$

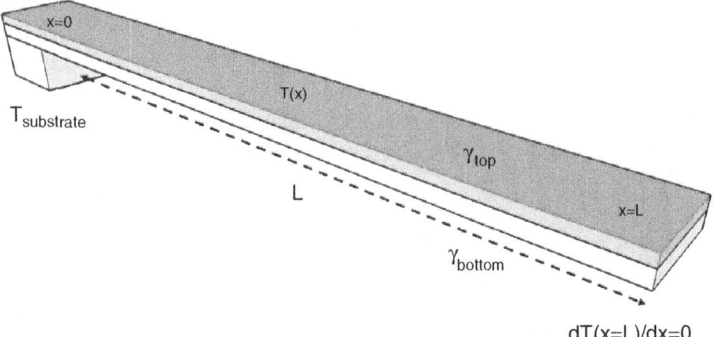

Figure 5.12 Bimorph cantilever beam actuator. The top layer of the cantilever has a coefficient of thermal expansion γ_{top} and the bottom layer has γ_{bottom}. One end of the beam, at $x = 0$, is thermally anchored to the substrate temperature, T_{sub}.

Applying the boundary condition that $dT/dx = 0$ at $x = L$ gives

$$\left.\frac{dT}{dx}\right|_{x=L} = -\frac{\tilde{P}_{Sources}}{\kappa}L + b = 0 \rightarrow b = \frac{\tilde{P}_{Sources}}{\kappa}L. \tag{5.39}$$

Integrating a second time,

$$\int \frac{dT}{dx}dx = T(x) = -\frac{\tilde{P}_{Sources}}{2\kappa}x^2 + \frac{\tilde{P}_{Sources}}{\kappa}Lx + c. \tag{5.40}$$

Applying the boundary condition $T(0) = T_{sub}$ gives $c = T_{sub}$:

$$T(x) = -\frac{\tilde{P}_{Sources}}{2\kappa}x^2 + \frac{\tilde{P}_{Sources}}{\kappa}Lx + T_{sub} \tag{5.41}$$

$$\Delta T(x) = -\frac{\tilde{P}_{Sources}}{2\kappa}x^2 + \frac{\tilde{P}_{Sources}}{\kappa}Lx. \tag{5.42}$$

This temperature distribution is plotted in Figure 5.13.

We can use this temperature distribution to find the deflection of a bimorph actuator. In one dimension the change in length of a cantilever beam would be

$$\Delta L = \int_0^L \gamma \Delta T(x) dx. \tag{5.43}$$

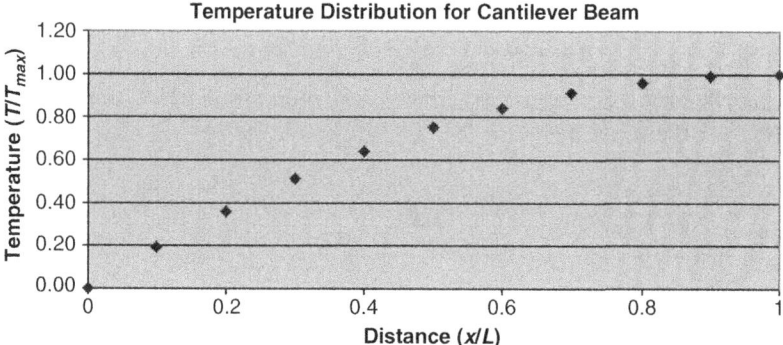

Figure 5.13 Temperature distribution for a fixed-free cantilever beam.

The strain ε would be given by

$$\varepsilon = \frac{\Delta L}{L} = \int_0^L \frac{\gamma \Delta T(x) dx}{L} = \frac{\gamma}{L} \int_0^L \Delta T(x) dx \qquad (5.44)$$

For a bimorph actuator comprised of two layers with different coefficients of thermal expansion, γ_1 and γ_2, the temperature change causes a strain mismatch between the layers, leading to curvature of the bimorph:

$$\varepsilon = \frac{\gamma_1 - \gamma_2}{L} \int_0^L \Delta T(x) dx. \qquad (5.45)$$

The curvature of the bimorph causes a deflection. If we simplify the analysis by considering two layers that only differ in their coefficients of thermal expansion, taking $E_1 = E_2 = E$ and $v_1 = v_2 = v$, and convert the strain to stress using the modulus of elasticity for a beam,

$$\sigma = \varepsilon \frac{E}{1 - v^2}. \qquad (5.46)$$

We can then use Stoney's formula to convert the stress into bending of the bimorph beam, similar to the bending of a substrate due to strain mismatch from a deposited film. Here the Young's modulus E_s, Poisson's ratio v_s, and the film thickness t_s are for the substrate, and the strain σ_f is for the deposited film with a thickness t_f:

$$\sigma_f = \frac{E_s}{1-v_s^2}\frac{t_s^2}{6t_f R} \to R = \frac{E_s}{(1-v_s^2)}\frac{t_s^2}{6t_f \sigma_f}. \qquad (5.47)$$

Using the approximation $d^2y(x)/dx^2 \approx 1/R$, integrating twice, and substituting for the stress σ,

$$y(x) = x^2/2R = \frac{x^2 \sigma_f (1-v_s^2)}{12 t_f E_s} = \frac{x^2(\gamma_1 - \gamma_2)}{12 t_f L}\int_0^L \Delta T(x)dx \qquad (5.48)$$

$$= \frac{x^2(\gamma_1 - \gamma_2)}{12 t_f L}\int_0^L \left(\frac{-\tilde{P}_{Sources}x^2}{2\kappa} + \frac{\tilde{P}_{Sources}Lx}{\kappa}\right)dx. \qquad (5.49)$$

For a bimorph made out of films with different thicknesses and Young's moduli, we can estimate the bending of the bimorph actuator using Stoney's formula modified for a bimorph:

$$\frac{1}{R} = \frac{(\gamma_2 - \gamma_1)\Delta T}{\frac{t}{2} + \frac{2(E_1 I_1 + E_2 I_2)}{t}\left(\frac{1}{E_1 a_1 b} + \frac{1}{E_2 a_2 b}\right)} \qquad (5.50)$$

$$I_1 = \frac{ba_1^3}{12} \quad I_2 = \frac{ba_2^3}{12}. \qquad (5.51)$$

Here γ_i are the coefficients of thermal expansion, ΔT is the temperature change above the ambient (where the bimorph is flat), E_i are the Young's moduli, I_i are the moments of inertia, a_i are the thicknesses of the two thin films that form the bimorph structure, and t is the total thickness of the films.

5.4 Bolometer

A bolometer is a thermally isolated, temperature-sensitive element that can be used to measure absorbed energy. Either surface micromachining or bulk silicon micromachining can be used to fabricate a structure with low thermal mass that is thermally isolated from the substrate by long, thin suspension arms. Here we consider a bolometer that is fabricated in the Poly2 layer of the PolyMUMPS process, as shown in Figure 5.14. If the device consists of a 100 (μm)^2 platform suspended by four suspension arms that are 5 μm wide, we can calculate the thermal conductance of the arms, the thermal capacitance of the platform, and the thermal time constant for the sensor.

5.4 Bolometer

Figure 5.14 Bolometer formed in Poly2 layer of a PolyMUMPS process. A serpentine resistor forms a suspended temperature sensor that is thermally isolated from the substrate.

The thermal conductance G of one of the arms is given by

$$G = \frac{\kappa A}{l} = \frac{(50 \text{ W/mK})(5 \times 10^{-6} \text{ m})(1.5 \times 10^{-6} \text{ m})}{100 \times 10^{-6} \text{ m}} = 3.75 \text{ μW/K}. \quad (5.52)$$

We have used an approximate value of 50 W/mK for the thermal conductivity of doped polysilicon at room temperature [2]. For the four arms the total thermal conductance would be 15 μW/K. The thermal conductance would increase if we include the metal wires that go to the temperature-sensitive element on the suspended platform. The thermal capacitance of the platform would be

$$C = \left(750 \frac{\text{J}}{\text{kg} \cdot \text{K}}\right)\left(2330 \frac{\text{kg}}{\text{m}^3}\right)(100 \times 10^{-6} \text{ m})^2 (1.5 \times 10^{-6} \text{ m})$$
$$= 2.62 \times 10^{-8} \text{ J/K}. \quad (5.53)$$

The thermal time constant for the bolometer would be given by the ratio of the thermal conductance to the thermal capacitance:

$$\tau = \frac{C}{G} = \frac{2.62 \times 10^{-8} \text{ J/K}}{15 \times 10^{-6} \text{ W/K}} = 1.75 \times 10^{-3} \text{ s}. \quad (5.54)$$

The bolometer would have a response time of a few milliseconds, enabling real-time infrared (IR) imaging for applications like night vision.

We can calculate the temperature change of the suspended platform when a single photon is absorbed. The energy of a single photon is given by

$$E = h\upsilon = \frac{hc}{\lambda}, \qquad (5.55)$$

where $h = 6.6261 \times 10^{-34}$ (J)(s) $= 4.1361 \times 10^{-15}$ (eV)(s) is Planck's constant, υ is the frequency of the photon in Hertz (Hz), $c = 2.99292458 \times 10^8$ m/s is the speed of light, and λ is the wavelength of the photon in meters. For a photon with wavelength $\lambda = 1.5\,\mu\text{m}$ in the IR, the energy of the photon would be 1.32×10^{-19} J, so the platform would be heated by

$$\Delta T = \frac{E}{C} = \frac{1.32 \times 10^{-19}\,\text{J}}{2.62 \times 10^{-8}\,\text{J}/\text{K}} = 5.05 \times 10^{-12}\,\text{K}. \qquad (5.56)$$

5.5 Thermal inkjet

One of the most commercially successful MEMS devices has been the thermal inkjet. In this device a small resistor heats an ink to create a vapor bubble. As the vapor bubble expands it displaces a drop of ink out of a nozzle. We will consider the design of a thermal inkjet device in Chapter 6 on microfluidics, but here we consider the heater that generates the vapor bubble. The typical propellant in inkjet inks is water, which is superheated to approximately 330°C in a few microseconds. If we consider a polysilicon heater that is $25\,\mu\text{m}$ on a side on top of oxide that is $2\,\mu\text{m}$ thick, the thermal conductance of the oxide would be

$$G = \frac{\kappa A}{l} = \frac{(1.25\,\text{W/mK})(25 \times 10^{-6}\,\text{m})^2}{2 \times 10^{-6}\,\text{m}} = 0.4\,\text{mW/K}. \qquad (5.57)$$

5.6 Thermal damage limits in thermally actuated MEMS

While thermal actuators provide a low-voltage, high-force means for actuation, there are some limitations associated with potential thermal damage mechanisms [3], [4]. These mechanisms include self-annealing and melting due to the formation of eutectics when silicon is in contact with metals such as gold. The standard process for polysilicon surface micromachining includes thermal anneal steps to reduce stress and stress-gradients. Typically these anneals occur at high temperatures, around 1050°C, and while the polysilicon is enclosed by phosphorsilicate glass (PSG), which acts as a solid source for the diffusion of phosphorous into polysilicon. When the thermal actuator is heated, the polysilicon is further annealed, causing mechanical changes in the polysilicon. Under sufficient heating the mechanical changes can cause permanent plastic deformations in the actuator, leading to nonrepeatable displacements that have been termed "back-bending." These deformations can be avoided by keeping the maximum temperature below approximately 800°C.

When silicon is in contact with a metal, a low-temperature liquid eutectic can be formed, far below the melting temperature of either silicon or the metal. For example, the eutectic temperature for silicon and gold occurs around 370°C for 30% silicon. In contrast, the melting point for pure gold is 1064°C, and the melting point for pure silicon is 1410°C [5]. Aluminum and silicon can form a eutectic around 577°C at around 12% silicon. In designing thermal actuators, such as gold/silicon thermal bimorph actuators or suspended heaters with metal wires and contacts, the metal should not be used in the proximity of any silicon that is heated unless the temperature is kept below the eutectic point.

Problems

(1) Consider the heatuator shown in Figure 5.4. We will make some simplifying assumptions to calculate the deflection of the actuator as shown in Figure 5.9. First, we will assume that the cold arm has a low electrical resistance in comparison to the hot arm, so that it has negligible Joule heating when a current is passed through the arm, and has a large thermal mass, so that it stays at the substrate temperature of 300 K for short heating pulses. We can then calculate the temperature increase of the hot arm and flexure using the heat diffusion equation for a fixed-fixed beam where the temperature at each end is fixed as the substrate temperature of 300 K (degrees Kelvin). Assume that the thermal conductivity of silicon is 1.5 W/cm-K at room temperature (300 K).

(a) For the hot arm assume the width $W_H = 2\,\mu m$ and the length $L_{HOT} = 200\,\mu m$, respectively. Calculate the average temperature of the hot arm for a power dissipation of 25 mW assuming the only heat loss mechanism is by conduction to the substrate and cold arm. Repeat the calculation to find the average temperature of the flexure. Assume the width $W_F = 2\,\mu m$ and the length $L_{FLEX} = 40\,\mu m$.

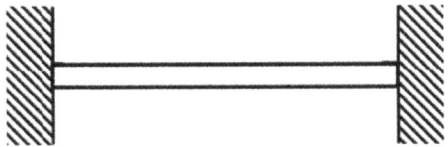

Figure 5.15 Top view of a fixed-fixed beam.

Assume the thickness of all the layers is 2 μm. Use the following formula for the average temperature:

$$T_{Average} = \frac{1}{L}\int_0^L T(x)dx.$$

(b) Using the formula below for a cantilever thermal bimorph, calculate the tip deflection δ, where T_{HOT} is the average temperature of the hot arm and T_{COLD} is the average temperature of the cold arm and flexure. Assume the cold arm remains at the substrate temperature of 300 K. L is the length of the hot arm (200 μm), and W_{H+F} is sum of the widths of the hot arm and flexure (2 μm + 2 μm = 4 μmh. λ is the coefficient of thermal expansion of polysilicon and is given as a function of temperature [K] by

$$\delta_{Lateral} = \frac{3L^2(T_{HOT} - T_{COLD})(\lambda(T_{HOT}) - \lambda(T_{COLD}))}{4W_{H+F}}$$

$$\lambda(T) = \left(3.725\{1 - e^{[-5.88\times 10^{-3}(T-124)]}\} + 5.548 \times 10^{-4}T\right) \times 10^{-6}\,[\text{K}^{-1}].$$

(2) A fixed-fixed polysilicon beam, as shown in Figure 5.15, is heated creating a compressive stress σ.
  (a) If the width of the beam is W, the length of the beam is L, and the thickness of the beam is t, find an expression for the compressive stress as a function of temperature. Assume the coefficient of thermal expansion for the beam is $\gamma = 2.3 \times 10^{-6}$/K and Young's modulus $E = 160$ GPa.
  (b) Once a critical value of stress is reached, the beam will buckle. This is called the Euler buckling limit and is given by

$$\sigma_{Euler} = -\frac{\pi^2}{3}\frac{Et^2}{L^2}.$$

5.6 Thermal damage limits in thermally actuated MEMS

At what temperature rise ΔT above ambient will a beam that is 100 μm long, 5 μm wide, and 2 μm thick buckle?

(c) If the beam is heated by passing a current through it, what current is required to buckle the beam? Assume the beam will buckle when the *average* temperature of the beam equals the ΔT found above. Also assume that the beam is fabricated in Poly1 of the PolyMUMPS process with a sheet resistance of 10 Ω/□ and that the thermal conductivity of polysilicon is $\kappa = 1.5$ W/cm-K. Ignore any variations in the parameters with temperature. The average temperature rise is given by

$$\Delta T_{Average} = \frac{1}{L}\int_0^L \Delta T(x)dx,$$

where $\Delta T(x)$ is given by

$$\Delta T(x) = T(x) - T_{substrate} = \Delta T = \left(\frac{\tilde{P}}{2\kappa}\right)x^2 + \left(\frac{\tilde{P}L}{2\kappa}\right)x.$$

(3) Describe how you can make a thermal actuator that can be used to position objects vertically out of plane. Can you find examples of this sort of actuator in the literature?

(4) Describe how you can make a thermal actuator for an inkjet printer in the PolyMUMPS process. To what temperature will you need to heat the water in the ink in order to quickly form a vapor bubble? How could you thermally isolate the heater from the substrate?

REFERENCES

1 Y.B. Gianchandani, *Bent-beam strain sensors*, J. Microelectromechanical Systems 5(1), pp. 52–58 (1996).
2 A.D. McConnell, S. Uma, and K.E. Goodson, *Thermal conductivity of doped polysilicon layers*, J. Microelectromechanical Systems 10(3), pp. 360–369 (2001).
3 J.H. Comtois and V.M. Bright, *Applications for surface-micromachined polysilicon thermal actuators and arrays*, Sensors and Actuators A: Physical 58(1), pp. 19–25 (1997).
4 J.H. Comtois and V.M. Bright, *Surface Micromachined Polysilicon Thermal Actuator Arrays and Applications*, Solid-State Sensor and Actuation Workshop, Hilton Head, pp. 174–177 (1996).
5 M. Hansen, *Constitution of Binary Alloys*, McGraw-Hill, p. 232 (1958).

6
Fluidic MEMS

Fluidic MEMS were some of the earliest and most commercially successful MEMS devices. The inkjet printer has displaced many of the other printing technologies for desktop and photographic color printing and is now penetrating the high-end digital printing market. An emerging market is developing for biological "lab-on-a-chip" and sensor applications. The same technology that enables printing color documents on a desktop may enable implantable medical devices to monitor internal chemical concentrations such as blood sugar levels and precisely and continuously dose drugs such as insulin on an as-needed basis. Before considering these applications we consider fluidics on micrometer length scales, as many of the phenomena we are used to on the macroscopic length scales, where our intuitions are formed, do not apply on the microscopic length scales of microfluidic devices.

6.1 Equations of motion

The equations of motion for fluids are given by the equation for the conservation of matter and Newton's law for the forces acting on the fluid. For an incompressible fluid, the density ρ of the fluid will be constant:

$$\frac{\partial \rho}{\partial t} = 0 \tag{6.1}$$

If the velocity of the fluid is $\vec{v}$, then the mass that flows in a unit time across a unit area of surface is the component of $\rho \vec{v}$ normal to the surface, so that the divergence of $\rho \vec{v}$ is equal to the decrease in density per unit time [1]:

$$\vec{\nabla} \cdot (\rho \vec{v}) = -\frac{\partial \rho}{\partial t} = 0 \rightarrow \vec{\nabla} \cdot \vec{v} = 0 \tag{6.2}$$

Newton's law for the forces acting on a unit volume of fluid with density ρ under the influence of a force f per unit volume can be written as [1]

$$\rho \vec{a} = \vec{f} \tag{6.3}$$

$$= -\vec{\nabla} p - \rho \vec{\nabla} \phi + \vec{f}_{visc} \tag{6.4}$$

where $-\vec{\nabla} p$ is the pressure force per unit volume, $-\rho \vec{\nabla} \phi$ is the force per unit volume associated with the potential ϕ, and $\vec{f}$ is the viscous force per unit volume. In lab-on-a-chip microfluidic devices, a common potential is an applied voltage to drive electrokinetic flow.

6.2 Microfluidics

Some of the important differences in the behavior of fluids on the microscale relative to the macroscopic scale we are used to in our everyday life are the following:

- All fluid flow is laminar rather than turbulent.
- Surface tension becomes an important force.
- Inertia becomes less important.
- The apparent viscosity increases.

Contrary to our intuition in the macroscopic world, fluids do not mix, bugs can walk on water, and our approach for swimming does not work. Understanding these differences is important to successful microfluidic design.

The relative importance of surface effects, like surface tension, to volume effects, like weight, can be understood through scaling laws. If we consider the surface area of a sphere, the surface area A varies with the radius r as

$$A = 4\pi r^2 \tag{6.5}$$

while the volume V of a sphere varies as

$$V = \frac{4}{3}\pi r^3 \tag{6.6}$$

so that the surface to volume ratio is given by

$$\frac{A}{V} = \frac{4\pi r^2}{4\pi r^3/3} = \frac{3}{r} \qquad (6.7)$$

As the radius decreases the ratio of surface to volume increases, so that surface forces such as surface tension and capillarity dominate forces that scale with volume like weight.

Viscosity is a distinguishing feature of fluids. As a fluid becomes more viscous, the less fluid-like it is. A related property in solids is the shear modulus or stiffness in shear. We previously found that the shear stress and shear strain in solids are related as follows:

$$\textit{Shear Stress} = (\text{Shear Modulus})(\textit{Shear Strain}) \qquad (6.8)$$

This can be written in the form

$$\textit{Stimulus} = (\text{Stimulus/Response})(\textit{Response}) \qquad (6.9)$$

For Newtonian fluids, the relationship between the stimulus and response can be written as

$$\textit{Shear stress} = (\text{Viscosity})(\textit{Shear Rate}) \qquad (6.10)$$

The main difference between solids and fluids is that a fluid will flow under the influence of a shear stress, whereas a solid will twist or deform. For fluids, the response to a shear stress is a shear rate with the proportionality constant that relates the stimulus and response given by the viscosity μ. The shear rate is a rate of change of velocity with distance, a velocity gradient, in the direction perpendicular to flow. It is the velocity gradient you feel when wading out into a creek from an eddy into the flow. A diagram illustrating how the viscosity μ is defined is shown in Figure 6.1. Two plates, each with area A, are separated by a distance d. One plate moves with velocity V under the influence of a force F. The velocity of the fluid varies linearly between the plates. At the lower stationary plate, the velocity of the fluid is zero, and it increases linearly, as shown by the length of the arrows, up to the maximum velocity V at the upper plate.

The viscosity μ is defined for a Newtonian fluid by the relation

$$\frac{F}{A} = \mu \frac{V}{d} \qquad (6.11)$$

The more viscous the fluid, the higher the force required to move the upper surface at velocity V. It is a measure of the resistance of fluids to flowing across surfaces or through conduits. The more viscous a fluid is,

6.2 Microfluidics

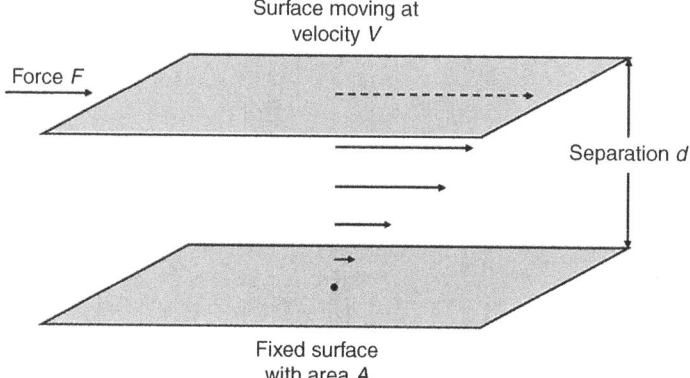

Figure 6.1 Definition of viscosity. Two plates, each with area A, are separated by a distance d. The lower plate is fixed. The upper plate moves with a velocity V under the influence of a force F. The velocity of the fluid varies linearly between the two plates. The steady viscous flow between the parallel plates, one of which is moving relative to the other, is called the Couette flow.

the less "fluid" it is. A familiar example for viscosity might be an ice cream maker, as shown in Figure 6.2. This device has two cylinders, one of which is stationary and the other of which can be rotated by a crank. As the cream turns into ice cream, the viscosity increases, making the crank harder to turn.

There are two types of fluid flows that can be important in microfluidic devices. One type of flow is called "Couette" flow, which is the steady viscous flow between parallel plates, one of which is moving relative to the other, as shown in Figure 6.1. The velocity of the fluid varies linearly from zero at the stationary plate up to the velocity V at the moving plate. Another type of fluid flow is "Poiseuille" flow. This is a pressure-driven flow between stationary parallel plates, as shown in Figure 6.3. It has a parabolic variation of the pressure with the maximum flow velocity in the middle of the plates and zero flow velocity at the walls.

6.2.1 Reynolds number

The Reynolds number, Re, is a dimensionless number that describes the relative importance of inertia and viscosity for the flow of a fluid. Considering Figure 6.1, the inertial and viscous forces would be

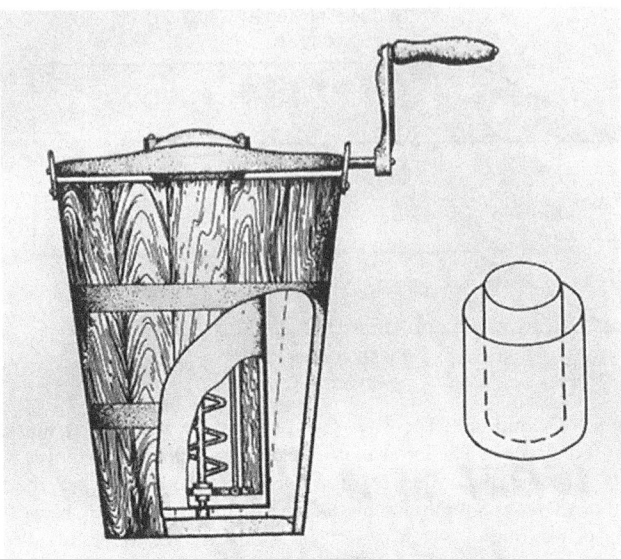

Figure 6.2 Ice cream maker. The handle rotates an inner cylinder. As cream between the rotating inner cylinder and the fixed outer cylinder turns into ice cream, its viscosity increases and the crank gets harder to turn. The inset shows the geometry of a viscometer for measuring fluid viscosity. It consists of two closely spaced cylinders, one of which is fixed and the other of which can rotate. With a narrow gap between the cylinders this geometry is similar to the plates shown in Figure 6.1.

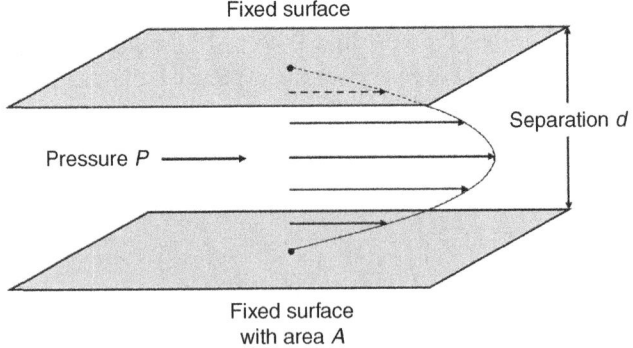

Figure 6.3 Poiseuille flow; a pressure-driven flow between two stationary, parallel plates.

6.2 Microfluidics

$$F_{inertia} = ma = \rho A d \frac{V}{t} = \rho A V^2 \tag{6.12}$$

$$F_{viscous} = \mu A \frac{V}{d} \tag{6.13}$$

so that the Reynolds number Re would be given by

$$\text{Re} = \frac{F_{inertia}}{F_{viscous}} = \frac{\rho A V^2}{\mu A \frac{V}{d}} = \frac{\rho d V}{\mu} \tag{6.14}$$

For high Reynolds numbers (Re > 2000) the inertial forces dominate the viscous forces, and the flow is turbulent. An example of turbulent flow would be a whale swimming; that would have a Reynolds number of around 200 000 000 [2]. In this turbulent regime there would be a wake with vortices in the water behind the whale. For low Reynolds numbers (Re < 1) the viscous forces dominate, and the flow is steady, or laminar. An example of laminar flow would be a paramecium swimming. In this laminar regime there would be no wake behind the paramecium, and the flow lines would be smooth and steady, with no formation of vortices. In microfluidics, the flow is usually laminar. In laminar flow conditions the mixing of fluids occurs by diffusion, which is slow compared to turbulent mixing.

6.2.2 Surface tension

As described in equation (6.7), as the length scale decreases, surface effects dominate volume effects, so that surface tension can dominate volume-related inertial forces. Bugs are able to walk on water as if the surface of the water were a trampoline. The surface of the water behaves like a stretched membrane due to the surface forces. Within the fluid all of the molecules are surrounded by other molecules, so they are fully bonded. At the surface of the fluid the molecules are bonded on one side, but not the other, so they are in a higher energy state. Minimization of the surface energy results in tension at the surface of the liquid to minimize the surface area. Surface tension causes a liquid drop to form a sphere to minimize the amount of surface area. To visualize the surface tension force, imagine making a cut in the stretched trampoline. The surface tension is the force F per unit length l that would have to be applied to hold together the two sides of the cut. The surface tension γ in Newtons/meter (N/m) is given by

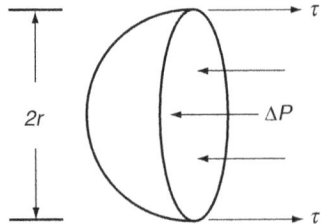

Figure 6.4 Surface tension for a meniscus in an inkjet nozzle with radius r. A pressure difference of ΔP from the fluid to the atmosphere causes a drop to bulge out of the orifice. Surface tension τ is exerted on the liquid for a length $2\pi r$ around the rim of the nozzle.

$$\gamma = \frac{F}{l} \qquad (6.15)$$

The direction of the surface tension is parallel to the surface and perpendicular to the cut. The surface tension can also be thought of as the energy in Joules per square meter (J/m^2) that is required to create a unit area of new surface. To create an additional amount of surface A, the required energy E would be

$$E = \int \gamma dA = \gamma A = \int \frac{F}{l} dA = \int F dx \qquad (6.16)$$

$$\gamma = \frac{E}{A} \qquad (6.17)$$

An example of surface tension is the formation of a meniscus at the orifice of a capillary filled with liquid, such as the nozzle of an inkjet print head. As shown in Figure 6.4, the liquid is held to the orifice by the surface tension τ around the periphery of the capillary. The pressure drop ΔP across the meniscus can be calculated by equating the force τ holding the meniscus to the capillary to the pressure pushing the fluid out of the capillary:

$$(2\pi r)\gamma = \Delta P(\pi r^2) \rightarrow \Delta P = \frac{2\gamma}{r} \qquad (6.18)$$

As a numerical example consider water ($\gamma = 7.27 \times 10^{-2}$ N/m) at a 10 μm nozzle as might be used in an inkjet printing device, as will be discussed later in this chapter. The pressure drop ΔP would be 1.45×10^4 N/m^2, or 0.14 atm.

6.2 Microfluidics

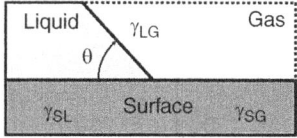

Figure 6.5 Contact angle. The interface between a solid, a liquid, and a gas. Surface tension γ_{SL} pulls to the left along the surface–liquid interface and surface tension γ_{SG} pulls to the right along the surface–gas interface. Surface tension γ_{LG} pulls in the direction of the liquid–gas interface and has a component of $\gamma_{LG} \cos(\theta)$ along the surface–liquid interface.

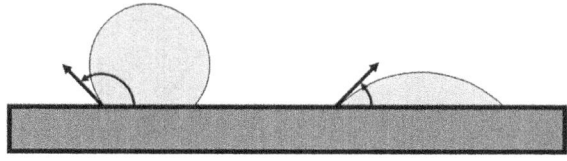

Figure 6.6 Contact angle. On a hydrophobic surface (left) the fluid balls up to decrease contact with the surface, forming a contact angle that is greater than 90 degrees. On a hydrophilic surface (right) the fluid spreads out to increase contact with the surface, forming a contact angle that is less than 90 degrees

6.2.3 Contact angle

Surface energy is the energy required to create a unit surface of a material. Interface energy is the energy required to create a unit of surface between two different materials. Both the surface energy and the interface energy determine the contact angle at the interface, as shown in Figure 6.5. Here there are interfaces between a gas, a liquid, and a solid. The surface tension γ is the interface energy per unit area or, equivalently, the interface force per unit length. The surface tension at the liquid–gas interface is γ_{LG}, at the solid–liquid interface it is γ_{SL}, and at the solid–gas interface it is γ_{SG}.

In equilibrium the liquid makes an angle θ with the solid surface, which is given by Young's equation:

$$\gamma_{SL} + \gamma_{LG} \cos(\theta) = \gamma_{SG} \tag{6.19}$$

If the angle is greater than 90 degrees, liquid balls up on the surface, as shown on the left-hand side of Figure 6.6, and the surface is said to be "hydrophobic," indicating a repulsive interaction between the liquid and the solid (water fearing). It balls up to avoid wetting the surface. If the contact angle is less than 90 degrees, the liquid wets the surface, as shown

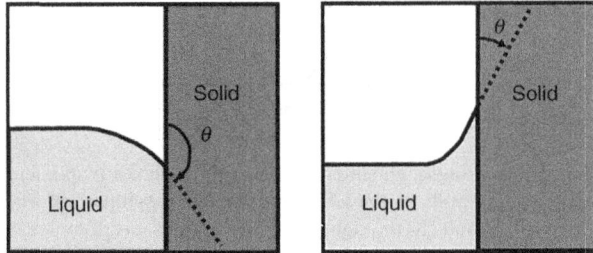

Figure 6.7 Contact angle at a vertical interface. Hydrophobic (90 degrees $< \theta <$ 180 degrees) (left) and hydrophilic ($0 < \theta <$ 90 degrees) (right).

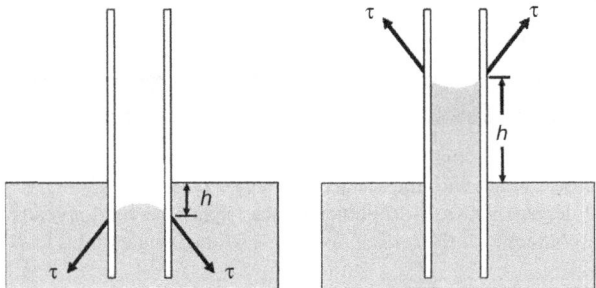

Figure 6.8 Capillary rise and depression. Hydrophobic (left) and hydrophilic (right).

on the right-hand side of Figure 6.6, and the surface is said to be "hydrophilic," indicating an attractive interaction between the liquid and the solid (water loving).

If the interface is vertical rather than horizontal, as shown in Figure 6.7, the force of gravity must be taken into consideration, leading to the phenomenon of "capillary rise."

6.2.4 Capillary rise

As shown in Figure 6.8, liquid in a small diameter tube, or capillary, will be pushed below the surface of the liquid outside the tube if the contact angle is greater than 90 degrees (hydrophobic), as shown on the left-hand side of Figure 6.8. The liquid will be drawn up into the tube to rise above the surface of the liquid outside the tube if the contact angle is less than 90 degrees (hydrophilic), as shown on the right-hand side of Figure 6.8.

The height of the capillary rise for a hydrophilic surface, or the depth of the capillary depression for a hydrophobic surface, can be found by equating the gravitational potential energy mgh with the energy due to the surface tension τ:

$$\rho g(\pi r^2 h) = 2\pi r \gamma \cos(\theta) \tag{6.20}$$

$$h = \frac{2\gamma \cos(\theta)}{\rho g r} \tag{6.21}$$

Here ρ is the density of the liquid that is lifted or pushed out of the tube, g is the gravitational constant (9.8 m/s^2), r is the radius of the tube, and h is the height of the liquid rise or the depth of the liquid depression.

6.3 Inkjet

Thermal inkjet is one of the leading commercial applications for fluidic microsystems. Its invention in the 1980s transformed the printing industry by bringing color to the desktop at low cost and high quality. In thermal inkjet printing a vapor bubble is quickly grown on top of a small resistive heater. A vapor bubble grows explosively as the water-based ink is superheated in microseconds to more than 300°C and displaces ink through a small orifice. Typically the diameter of the orifice is equal to the diameter of the ink drop, which can be from 10 to 20 μm, depending on the resolution of the printer.

One of the challenges of thermal inkjet printers is the heat that is generated and the power that is required. The waste heat has to be removed and the amount of power required for resistive heating can be high, particularly for high-speed printing, which requires high-frequency drop ejection from multiple drop ejectors. The energy E required to eject a drop of ink is approximately 1 μJ, and the drop ejection frequency f can be as high as 10 kHz, resulting in a power dissipation for each jet of

$$P = \frac{\Delta E}{\Delta t} = \Delta E f = (1 \times 10^{-6} \text{J})(10 \times 10^3 \text{ drops/s}) = 10 \text{ mW} \tag{6.22}$$

Current inkjet printers have hundreds of nozzles, and future inkjet printers are projected to have several thousands of nozzles in page-wide arrays to obtain very high print speeds, which will require very high levels of power. Here we consider nonthermal means of drop ejections to decrease the required power.

One potential solution is to use piezoelectric or electrostatic actuation. Both of these approaches are mechanical rather than thermal means of

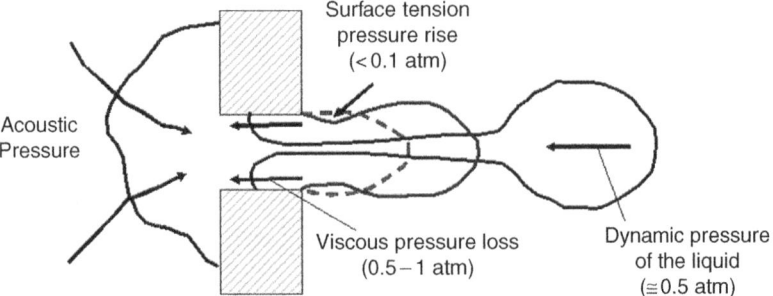

Figure 6.9 Pressure required to form and eject an ink drop in inkjet printing. The acoustic pressure generated by an actuator must overcome the viscous pressure drop required to force liquid through the nozzle (frictional energy loss), the surface tension pressure rise to form the drop (surface energy), and the dynamic pressure (kinetic energy) of the liquid.

actuation, so they have potential for lowering the required power. Piezoelectric actuators do not have high amounts of displacement, so drop ejectors based on piezoelectrics tend to be large (mm) and do not scale down to the 20–40 μm pitch size required for high-resolution printing at 600–1200 dots per inch. Here we consider two different types of electrostatic actuators that have been fabricated in multiwafer processes. The first drop ejector has been fabricated in the PolyMUMPS process and uses a hybrid packaging approach, where a nozzle plate is bonded onto a surface micromachined actuator. The second approach uses a monolithic packaging approach, where the nozzle plate is fabricated as a part of the surface micromachining process.

The acoustic pressure that is required to eject an ink droplet is shown schematically in Figure 6.9. To eject an ink drop, the acoustic pressure that is generated by the electrostatic actuator must overcome the viscous pressure drop required to push the ink through the nozzle (∼0.5–1 atm), the capillary pressure drop due to surface tension (∼0.1 atm), and the dynamic pressure of the liquid associated with its kinetic energy (∼0.5 atm). The total acoustic pressure required to eject a drop will be on the order of 1–2 atm of pressure.

A schematic cross-section diagram of a drop ejector fabricated in the PolyMUMPS process is shown in Figure 6.10. The actuator is a clamped circular diaphragm similar to the M-Test structure CD discussed in Chapter 2, Section 2.8. The membrane diaphragm is defined in the Poly1 layer (2 μm) and is pulled down by application of a voltage to a

6.3 Inkjet

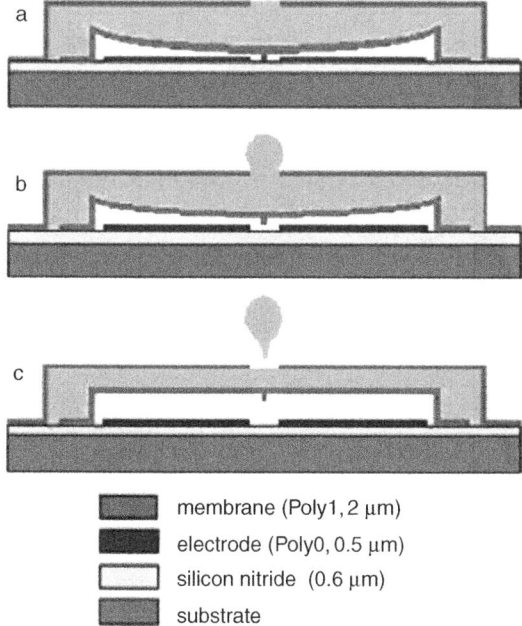

- membrane (Poly1, 2 μm)
- electrode (Poly0, 0.5 μm)
- silicon nitride (0.6 μm)
- substrate

Figure 6.10 Electrostatic membrane drop ejector. (a) The membrane has been pulled down by the application of a voltage applied to the counter-electrode. (b) The voltage is removed and the membrane relaxes back to its initial position, ejecting a drop of ink through the orifice in the nozzle plate, as shown in (c). The membrane is fabricated in Poly1 and the counter-electrode is fabricated in Poly0. The nozzle plate with the orifice can be fabricated in Poly2 or in a separate thick polymer layer such as SU-8 (Xerox MEMSJet). See color plate section.

counter-electrode defined in the Poly0 layer (0.5 μm), as shown in Figure 6.10(a). The Poly1 membrane, which is exposed to the ink, is held at ground potential to avoid exposing the ink to the voltage, which could lead to the electrolysis of a conducting ink. The gap between the counter-electrode and the diaphragm is determined by the 2 μm thick sacrificial Oxide1. Alternatively, for a larger gap, to generate a larger drop size, the diaphragm can be formed in the Poly2 layer and Oxide1 (2 μm) and 2 (0.75 μm) stacked to form a 2.75 μm gap. To eject a drop, the voltage is removed and the membrane relaxes back to its initial position, as shown in Figure 6.10(b) and (c). The ink that was located between the diaphragm and the nozzle plate is ejected through the orifice in the nozzle plate. The nozzle plate can be defined in Poly2 of the PolyMUMPS process or by a separate thick polymer such as SU-8, a thick photosensitive epoxy that is deposited and patterned as described in Chapter 7.

If the nozzle plate is fabricated in Poly2, the fluidic design will have to comprehend the flexure of the 1.5 μm thick nozzle plate under the applied pressure of the ink as the thicker Poly1 membrane (2.0 μm) is relaxed.

We can estimate the pressure that the membrane is able to generate by the pressure that is required to pull it down to the counter-electrode. If we use the parallel plate approximation for the electrostatic force found in Chapter 3, equation (3.13), the pressure is given by

$$P = \frac{F}{A} = \frac{\varepsilon_0}{2}\left(\frac{V}{g}\right)^2 \qquad (6.23)$$

where V is the applied voltage and g is the gap. For a 1 μm gap, the voltage required to generate a pressure of 2 atm would be

$$V = g\sqrt{\frac{2P}{\varepsilon_0}} = (1 \times 10^{-6}\,\text{m})\sqrt{\frac{(2\,\text{atm})(101,325\,\text{Pa/atm})}{8.85 \times 10^{-12}\,\text{F/m}}} = 150\,\text{V} \qquad (6.24)$$

If the membrane is actuated under voltage control, it will actually pull in at a lower voltage, once the original gap has been decreased by a little more than one-third of its initial spacing, as shown previously in equation (3.24), or about 0.7 μm. The gap can be closed a little further than one-third of the initial voltage because the clamped boundary conditions lead to a nonlinear spring as the membrane stretches when it is displaced by an amount more than its thickness.

An image of an actuator that has been pulled down is shown in Figure 6.11. Here the deflection of the membrane can be seen by the formation of Newton's rings when a coherent source is used for illumination.

Light from the counter-electrode interferes with the light reflected from the membrane, giving an interference ring each time the optical path length changes by half a wavelength:

$$\ell = m\frac{\lambda}{2} \qquad (6.25)$$

In this case the membrane was illuminated by red light from a helium neon laser that had a wavelength of 0.63 μm, so three dark rings are formed at 0.32 μm, 0.63 μm, and 0.95 μm gap spacings.

A stroboscopic image of drops being ejected is shown in Figure 6.12. The nozzle is indicated by a dashed line. The reflection of the drop that is being ejected can be seen in the nozzle plate to the left in the figure. Drops that have been ejected can be seen on the right.

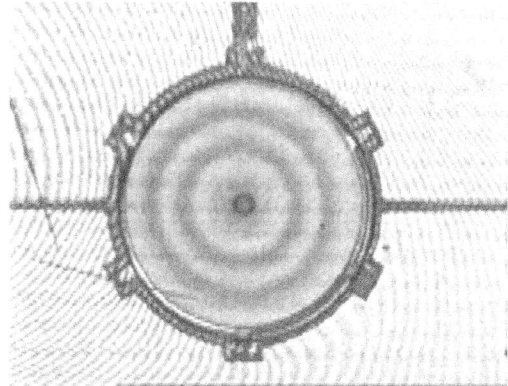

Figure 6.11 Formation of Newton's rings for a clamped-circular diaphragm defined in Poly1 that has been pulled down by the application of a voltage to Poly0 (Xerox MEMSJet).

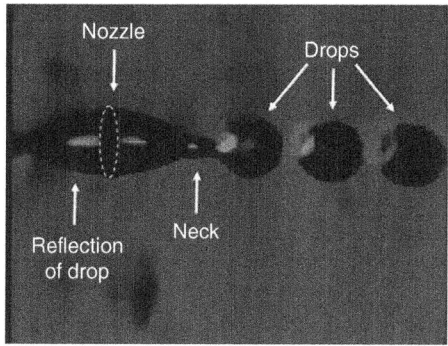

Figure 6.12 Drop ejection from a MEMSJet drop ejector (Xerox MEMSJet).

A second type of inkjet with an electrostatic actuator has been fabricated in the Sandia SUMMiT V process [3], [4]. In this design, the ink fills in the gap in the electrostatic actuator, increasing the dielectric constant and thus the pressure that is generated. Since the ink is exposed to the drive voltages, a high-frequency RF drive signal is used to avoid problems with electrolysis of the ink. A schematic cross section of the drop ejector is shown in Figure 6.13 [5].

A scanning electron microscope (SEM) image of a piston drop ejector with part of the front face removed to show the actuators is included in Figure 6.14.

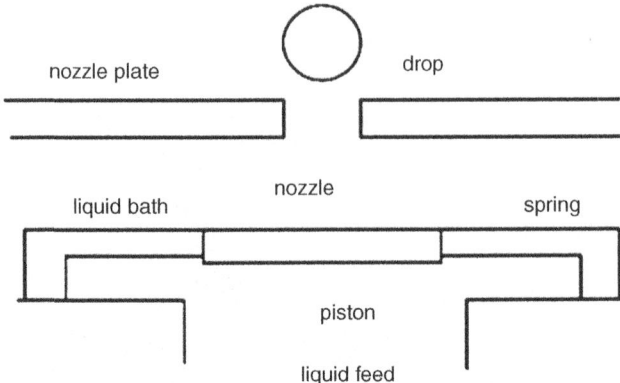

Figure 6.13 Electrostatic piston drop ejector. (Reprinted with permission from *A Surface Micromachined Electrostatic Drop Ejector*, P. Galambos, K. Zavadil, R. Givler, F. Peter, A. Gooray, G. Roller, and J. Crowley, ©1967 IEEE.)

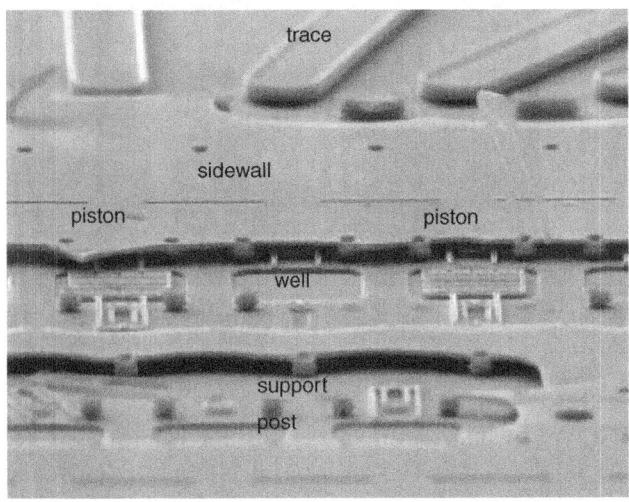

Figure 6.14 Cut-away SEM image of a piston drop ejector that was fabricated in the SUMMiT V surface micromachining process. (Reprinted with permission from *A Surface Micromachined Electrostatic Drop Ejector*, P. Galambos, K. Zavadil, R. Givler, F. Peter, A. Gooray, G. Roller, and J. Crowley, ©1967 IEEE.)

6.3 Inkjet

Problems

(1) Explain why a steel needle can float on water while a steel nail cannot. Assume water has a surface tension of 72×10^{-3} N/m and has a contact angle of 30 degrees with either the needle or nail. Assume the needle has a length of 2 cm and a diameter of 1 mm. Assume the nail has a length of 4 cm and a diameter of 4 mm. The density of steel is 8 g/cm^3.

(2) For each of the coins in your currency, determine whether the coin will float on water.

(3) Design a MEMS inkjet drop ejector in the PolyMUMPS process, as shown in the cross section in Figure 6.10. You do not need to worry about the fluid reservoir shown above the actuator that is formed as a circular diaphragm in Poly1 (red).

  (a) Assume that you want to print at a resolution of 600 dpi (dots per inch). An ink droplet will typically spread out into a dot approximately twice its diameter when it hits the page. Calculate the required drop volume for printing at 600 dpi resolution (1 in. = 25.4 mm).

  (b) For a circular diaphragm with radius r fabricated in Poly1, *estimate* the volume displacement expected if the membrane is pulled down a distance δ_{center} until a dimple on the bottom of Poly1 lands on the silicon nitride dielectric layer. Assume the dimple has a height of 0.75 μm and that sacrificial Oxide1 is 2 μm thick. What radius is required to generate the appropriate sized drop volume for 600 dpi printing?

  (c) Assume the central deflection of a circular membrane is given by

  $$\delta_{center} = \frac{3Pr^4(1-v^2)}{16Et^3}$$

  where $v = 0.20$ is Poisson's ratio and $E = 160$ GPa is Young's modulus. Calculate the pressure that is required to obtain the center deflection used in part (b) above (e.g., the membrane is pulled down until a 0.75 μm dimple lands on the nitride dielectric layer).

  (d) What voltage would be required to generate this pressure? Assume that the pressure $P = F/A$, where F is the electrostatic force (e.g., assume the bending of the membrane is negligible and that you can use the equation for the electrostatic force for a

parallel plate actuator). The permittivity of free space $\varepsilon_0 = 8.85 \times 10^{-12}$ F/m.
  (e) To eject a drop, the actuator must overcome the pressure to create a meniscus. Assuming the nozzle is 20 μm in diameter, what is the required pressure to generate the meniscus prior to drop ejection?
(4) Design a fluidic pump in a multiproject wafer process. The pump should have a volumetric flow rate of 20 nL/s for water through a 100 μm × 100 μm microchannel that is 1 mm long. What pressure difference is required to achieve this flow rate? Before starting, perform a literature search to see what has been done in the past.
  (a) Design a displacement pump (e.g., a moving piston that displaces fluid).
  (b) Design a rotary pump (e.g., using a turnstile mechanism).
  (c) Design a peristaltic pump (e.g., a sequenced deformation of a channel or a sequenced formation of vapor bubbles).
  (d) Design a capillary pump (e.g., the fluid flow is driven by capillary forces).
  (e) Design an electrokinetic pump (e.g., using electrophoresis or electro-osmosis).

REFERENCES

1 Richard P. Feynman, Robert B. Leighton, and Matthew Sands, *The Feynman Lectures on Physics, Volume II*, Addison-Wesley Publishing Company, Reading, MA, pp. 40–2 to 40–3 (1964).
2 S. Vogel, *Life's Devices, the Physical World of Animals and Plants*, Princeton University Press, Princeton, NJ, pp. 113–117 (1988).
3 A. Gooray, G. Roller, P. Galambos, K. Zavadll, R. Givler, F. Peter, and J. Crowley, *Design of a MEMS ejector, for printing applications*, J. Imaging Science and Technology 46(5), pp. 415–432 (2002).
4 E.P. Furlani, *Analysis of an electrostatic MEMS squeeze-film drop ejector*, Sensors & Transducers J. 7, Special Issue, October, pp. 78–87 (2009).
5 P. Galambos, K. Zavadil, R. Givler, F. Peter, A. Gooray, G, Roller, and J. Crowley, *A Surface Micromachined Electrostatic Drop Ejector*, Transducers'01, June 2001.

7
Package and Test

In a multi-project wafer process your layout will be included with other users' layouts that are combined together into a wafer level layout. All of the designs will be fabricated together using the same process. When all of the fabrication steps have been completed the wafer is subdivided into individual dies for each of the users using an abrasive diamond saw. Dicing "streets" are put in at the wafer level layout by the supplier to comprehend the "kerf" (width) of the saw blade and a safety zone for chipping during the sawing operation. The saw blade and wafer are sprayed with water to keep them cool during the dicing operation, and the sawing process generates residual particles that must be removed in a post-sawing cleaning step. At this stage custom steps can be performed for each individual customer at the die level. Some of the custom steps include sub-dicing of the individual die into smaller pieces, post-processing steps such as sacrificial release and critical point drying, and packaging of the parts. Packaging can include attaching the die to a carrier, wire bonding and sealing the package to protect the parts from the environment and to provide a controlled interface (electrical, thermal, mechanical, and/or optical) between the parts and the environment.

7.1 Release

A typical process sequence for the fabrication of a MEMS device is shown in Figure 1.4. At the end of the fabrication cycle the MEMS part is released from the substrate by etching a sacrificial layer in a surface micromachining process, or by etching selected regions of the substrate in a bulk micromachining process. This enables a micromechanical element such as a proof-mass or an actuator to move independently of the substrate, or a sensor element such as a temperature sensor to be thermally isolated from the substrate.

Prior to release the MEMS dies typically look as expected by the designer. After release the released parts can deform under the action of the residual stress and stress-gradients, and parts that are not attached with an anchor can float away. Examples of released parts that have deformed after release are shown in Figure 1.6 and bond pads that were not properly anchored and have become detached from the substrate are shown in Figure 1.9.

Many of the suppliers for multiproject wafer processes will offer services for release of the parts at the die level. If this service is offered it is highly recommended for users who do not have access to a cleanroom for performing their own release. The release steps typically use hazardous chemicals such as hydrofluoric (HF) acid in a polysilicon surface micromachining process, or potassium hydroxide (KOH) in a bulk micromachining process, which require the proper chemical handling and safety equipment. Hydrofluoric acid is particularly insidious as it absorbs through the skin before any burning sensation is felt. The supplier may also offer advanced release procedures such as a dry release to avoid stiction effects, as shown schematically in Figure 1.10. An example is the use of a HF vapor instead of liquid HF to avoid the formation of a liquid meniscus that can cause stiction from the capillary forces as the part is dried. Another approach is to avoid the formation of a liquid–vapor interface altogether by using a supercritical drying process [1]. This is accomplished by controlling the pressure and temperature of the drying process, typically using liquid carbon dioxide (CO_2) that is brought into a supercritical state critical point (304.25 K at 7.39 MPa or 31.1 degrees Celsius at 1072 psi) to avoid crossing the phase boundary between a liquid and gas where a meniscus would be formed.

A final approach to avoid stiction effects is to apply a self-assembled monolayer (SAM) that is hydrophobic as a part of the release process. This process forms a hydrophobic Teflon-like coating on the contacting surfaces that reduces the capillary forces decreasing stiction effects [2].

After the parts have been released they can be inspected to determine whether or not the release went well. If parts have not been properly anchored the die can look like a battleground with debris distributed over it. A good design practice is to include test structures in addition to the parts to assess the success of the release step. Common test structures include arrays of different length cantilever and fixed-fixed beams, as discussed in Chapter 2. The cantilever beams can be inspected to see whether or not they were properly released without stiction. If there are stiction effects then the critical length beyond which the beams are too flexible and stiction occurred can be determined and comprehended in future designs. The cantilever beams will also indicate any residual stress gradients. If there are residual stress gradients the beams will deflect as shown in Figure 1.5. If the amount of vertical deflection can be measured as a function of beam length it can be used to predict deflections from residual stress gradients in future designs. Arrays of fixed-fixed beams can be used to detect residual compressive stress by determining the critical length for Euler buckling, as discussed in Chapter 2. Finally, bent-beam strain sensors oriented parallel to the die edges can be included to determine the magnitude of compressive or tensile residual stress, as shown in Figure 2.16.

7.2 Test equipment

One of the primary pieces of test equipment is a probe station that allows mechanical and electrical probing of MEMS parts (Figure 7.1). The probe station will have a microscope with a number of different objective lenses, allowing different levels of magnification and fields of view. A 1–3× zoom feature is also useful. Ideally the objective lenses will have a long working distance to enable changing the objective without having to increase the lens–sample spacing and to allow bringing in the probes below the objective lens to contact the sample.

The probe station should have some means for holding down the device under test. A common approach is to use vacuum hold down, but many probe stations have platens with vacuum ports spaced to hold down whole wafers rather than diced parts. It is possible to block some of the outlying vacuum ports with tape so that sufficient vacuum can be developed in the inner vacuum ports to hold down the die. Alternatively, the die can be held in place with double-sided sticky tape, but this tape can leave a residue on the probe station platen and the back of the part under test.

The electrical testing described requires landing a probe onto the bond pad. This can be fairly straightforward for layouts with large bond pads (≥ 150 μm square) or as challenging as landing a jet on an aircraft carrier for very small bond pads or for bond pads that are spaced closely

Figure 7.1 A probe station is used for electrical and mechanical probing of the MEMS parts. A microscope with objectives that have a long working distance allows micromanipulators to contact the device under test. It is useful to have a range of magnifications that allow imaging of both the entire chip, to find particular devices on the chip, and imaging of the device of interest. Electrical signals can be brought to the chip to actuate it, such as the voltage for an electrostatic actuator, and brought from the chip to detect signals from it, such as the change in resistance of a piezoresistive sensor. The probes can also be used to apply a mechanical force to make sure that parts that are supposed to be released actually are released. Bond pads on the MEMS device under test should be large enough to easily land the probes on them, on the order of 100 to 150 μm on a side. The landing of the probe can be detected as a horizontal deflection of the probe when contact is made.

together. A first step is to get the probes within striking distance of the bond pad without destroying any of the parts on the chip. The probes are lifted sufficiently high that they will not contact the chip, but sufficiently low that they are visible within the working distance of the probe stations optics. A good option is to use a microscope objective that has sufficient depth of focus to enable seeing both the probe and the bond pad during the landing. It is useful to start off with low-power objectives that allow a wide field of view as the probe is brought towards the bond pad, and then to switch to a higher power objective with a smaller field of view when the probe has been brought into close proximity of the bond pad. Once the probe is close to the pad it can be lowered down to the pad to be brought into contact. When the probe comes into contact with the pad it will deflect sideways and can sometimes leave a "witness" mark or scratch on the pad as it slides across it.

For designs that require making multiple electrical connections to bond pads for their operation, it is good practice to limit the number of bond pads by making on-chip electrical connections with wires rather than off-chip connections through bond pads. It can be very difficult bringing multiple probes into contact with the chip without encountering either mechanical interference between the probes, or optical interference that limits the field of view. If multiple off-chip connections are required the chip can be mounted on an interposer board with fan-out to bond pads that are more widely spaced. If a common pad layout is re-used a "probe-card" can be designed with that layout that allows multiple probe contacts to be made at the same time.

7.3 Mechanical testing

Often parts do not perform as expected, so it can be useful to include some simpler mechanical test structures to allow probing whether or not parts have been released as expected. A useful test structure is a spring that is anchored at one end, with a ring at the other end that is large enough to fit a probe into. A ring diameter of 150 μm should be sufficient. The probe can be lowered into the released ring, and moved laterally to ensure that the released spring moves as expected.

If a released part does not move as expected it is possible to poke the part with the probe, often times without breaking the part. If the release step was not carried through to completion, sometimes the remaining residual sacrificial films that have been partially etched can be broken by mechanical probing without destroying the part that is to be released. Also, if the part moves "sluggishly" it can be an indication that there is a residual contaminant film or absorbed water layers between the released part and the substrate if the released parts have not been stored in a dry environment such as a storage cabinet with desiccant.

7.4 Electrical testing

In addition to mechanical test structure it is good practice to include electrical test structures. Simple structures like electrical bond pads that are connected together electrically by wires defined in different layers can be tested to ensure the electrical continuity that is expected. This sort of information can be very valuable in problem solving in case designs do not work as planned. The contact resistance can then be measured to ensure good electrical contact. If the contact resistance is high the probe can be removed for cleaning. If cleaning the probe does not improve the contact resistance there may be a problem with the bond pad design or layout such as a residual insulating layer that was not removed to permit electrical conductivity.

If the die substrate must be held at ground potential for the electrical testing, it will be necessary either to include a ground pad that is electrically connected to the substrate, or to make contact to the substrate through the back of the die. In order to make a good backside contact, all of the electrically insulating layers will need to be removed from the backside. One way to accomplish this is to scratch off any backside films and to use electrically conducting glue such as "aquadag," a solution with a colloidal suspension of conducting graphite particles, to hold the chip down. This conducting glue is often used in scanning electron microscopes to avoid charge buildup on parts that are imaged with the electron beam. If the part under test is held down on the probe station platen by double-sided sticky tape, a backside ground contact will require making contact at the edge of the die to short circuit the insulating tape.

As described in Chapter 2, the M-Test structures are useful for determining thin film properties as well as testing electrical functionality. By applying increasing voltages to the M-Test structures, they can be optically inspected to determine the voltage that is required for pull-in. These simple test structures enable a rapid assessment to determine quickly if released parts are able to be actuated as expected using electrostatic forces.

Often the supplier will include one of their own test structures in the customer's layout space. For example the PolyMUMPs die will have a comb-drive resonator added into the layout after the customer turns it in, so long as they have followed the design guidelines that specify leaving a space on the layout for the inclusion of the comb-drive resonator. The comb-drive can be inspected for the fidelity of the patterning since the combs of the comb-drive use the minimum space to achieve the maximum force. The comb-drive can also be driven into electrical resonance by the application of the appropriate biasing conditions [3].

7.5 Optical characterization

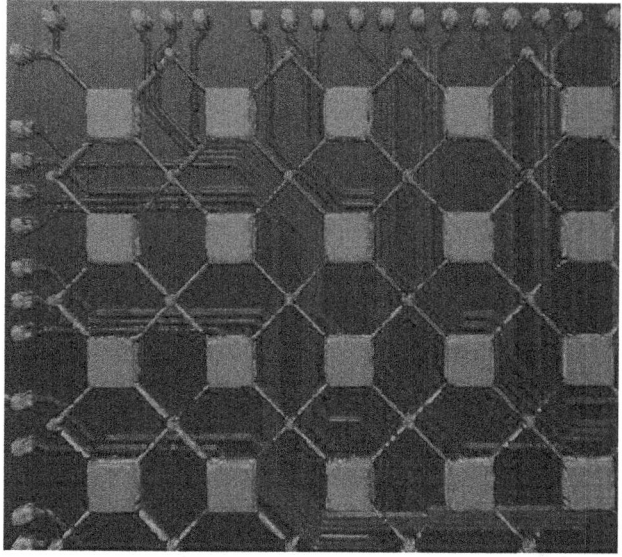

Figure 7.2 A white light interferometer can be used to make measurements of the device under test without contacting it. This technique is most often used to obtain height information for MEMS structures. It can be useful to measure out-of-plane displacement information for vertical actuators. See color plate section.

7.5 Optical characterization

Often it is possible to detect in-plane motion of MEMS devices by observing the part as it is actuated. If the motion is small the part can be excited into resonance by applying an alternating current (ac) signal at the parts resonance frequency where the motion will be amplified for a high-Q structure. If the vibration frequency is too high to follow, often the envelope of the motion can be observed. For out-of-plane motion a high-magnification objective with a short working distance can be used. As the part moves out of plane the focus will be lost. By raising the objective to bring the part back into focus, the out-of-plane deflection can be measured so long as the z-travel for the microscope is calibrated.

A very useful tool is a white light non-contact interferometer that can be used to obtain three-dimensional renderings of the part. It can also be used to extract z-height data and surface roughness. A white light interference image of an array of X-beam electrostatic actuators is shown in Figure 7.2. The different colors correspond to different z-heights of the actuators. When a voltage is applied to the actuator the central square

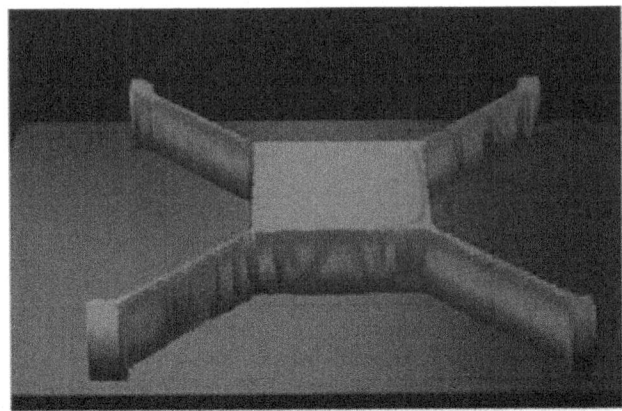

Figure 7.3 Displacement of an individual X-beam actuator. (Top) Undeflected actuator. (Middle) Partial deflection. (Bottom) Maximum deflection. See color plate section.

7.5 Optical characterization

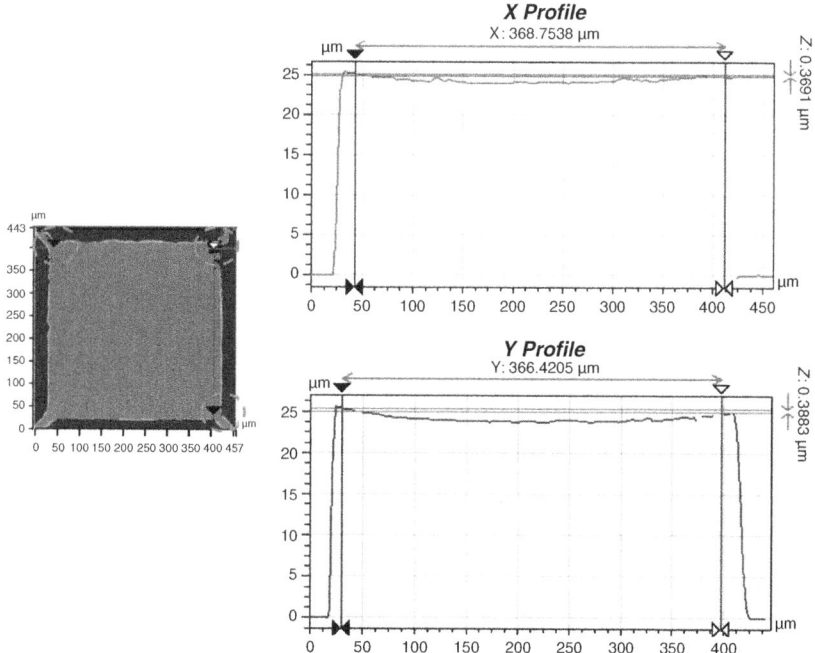

Figure 7.4 Line scans across a MEMS actuator taken with white light noncontact interferometry. Line scans can be used to make quantitative measurements of out-of-plane dimensions.

is pulled downwards towards the substrate. The displacement of an individual actuator is shown in Figure 7.3. A line scan can be used to measure the deflection along the line as shown in Figure 7.4. Stroboscopic imaging can also be used to analyze MEMS motion [4], [5].

REFERENCES

1. Ijaz Jafri, Heinz Busta, and Steven Walsh, *Critical point drying and cleaning for MEMS technology*, Proceedings SPIE 3880, pp. 51–58 (1999).
2. Roya Mubovidian, W. Robert Ashurst, and Carlo Carraro, *Self-assembled monolayers as anti-stiction coatings for MEMS: Characteristics and recent developments*, Sensors and Actuators 82, pp. 219–223 (2000).
3. William C. Tang, Tu-Cuong H. Nguyen, and Roger T. Howe, *Laterally Driven Polysilicon Resonant Microstructures*, Micro Electro Mechanical Systems, IEEE Proceedings Vol. 20–22, pp. 53–59 (1989).
4. Christian Rembe, Rishi Kant, and Richard S. Muller, *Optical measurement methods to study dynamic behavior in MEMS*, Proceedings of SPIE 4400, pp. 127–137 (2001).
5. Tong Guo, Hong Chang, Jinping Chen, Xing Fu, and Xiaotang Hu, *Micro-motion analyzer used for dynamic MEMS characterization*, Optics and Lasers in Engineering 47, pp. 512–517 (2009).

8
From Prototype to Product: MEMS Deformable Mirrors for Adaptive Optics

MEMS deformable mirrors (DMs) have been developed for applications in adaptive optics, including astronomy [1], [2], [3], vision science [4], microscopy [5], and laser communications [6]. In astronomy, adaptive optics have been used to overcome the image aberrations caused by the Earth's atmosphere. Light from a distant star, which can be considered a point source because it is so far away, travels through the vacuum of space as a plane wave. When the plane wave enters Earth's atmosphere, the wavefront is distorted due to dynamic changes in the index of refraction of the atmosphere caused by winds and temperature fluctuations. These fluctuations in the index of refraction cause changes in the velocity of the wavefront, so that some portions travel faster than others, leading to the distorted wavefront shown in Figure 8.1. These dynamic distortions are what cause stars to appear to twinkle. When the star is imaged in a telescope, it appears as a fuzzy blob rather than a point of light, as shown in Figure 8.1(a). By measuring the wavefront distortions from the star using a wavefront sensor, the conjugate of the wavefront distortion can be applied to a deformable mirror to correct the image, as shown in Figure 8.1(b) and (c). When a star is used as a reference point source for making wavefront corrections, it is called a "guide-star." If light from a nearby galaxy travels through the same part of the atmosphere, the guide-star can be used to correct the image of the galaxy, as shown in Figure 8.1(b) [7].

Examples of the use of adaptive optics in astronomy and vision science are shown in Figure 8.2. An image of the planet Neptune is shown in Figure 8.2(a) with conventional optics and corrected with adaptive optics [8]. Images of clouds on the planet can be seen in the corrected image. An uncorrected and corrected image of a living human retina is shown in

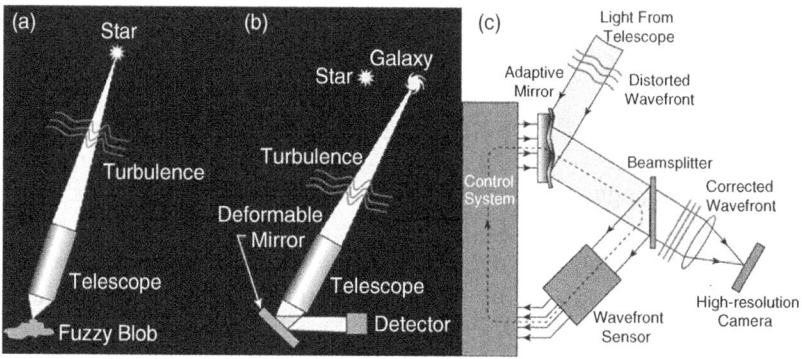

Figure 8.1 Adaptive optics in astronomy. (a) The image of a star appears as a fuzzy blob in a telescope due to the wavefront aberrations caused by turbulence in the atmosphere. (b) If a deformable mirror is used to correct the image of the star to make it back into a point of light, a nearby galaxy with a more complex structure will also be corrected. (c) The wavefront distortions can be measured with a wavefront sensor. The wavefront distortions from the "guide star" are fed back through a control system to deform the adaptive mirror to a shape that is conjugate to the distorted wavefront, correcting the reflected wavefront of the "guide star" back into a plane wave. (Credit: Claire Max, Center for Adaptive Optics.) [7].

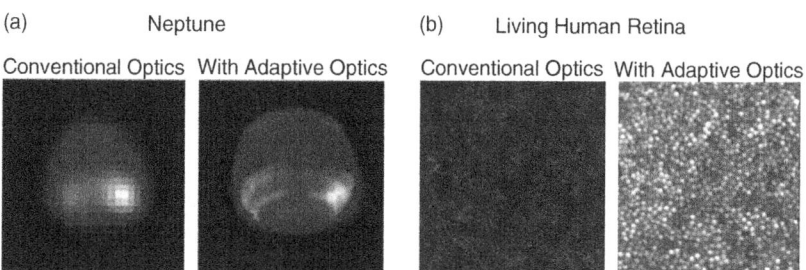

Figure 8.2 The use of adaptive optics to correct image aberrations. (a) Astronomy: Neptune observed in the near-IR (1.65 µm) with and without adaptive optics. (Credit: C.E. Max et al., reproduced by permission of the America Astronomical Society). (b) Vision science: imaging of individual rods and cones in the living human retina. (Credit: Y. Zhang, S. Poonja, and A. Roorda, reproduced by permission of the Optical Society of America.) See color plate section.

Figure 8.2(b). The mosaic of the individual rods and cones can be seen in the corrected image [9].

Adaptive optic technology in the past has made use of piezoelectric actuators to deform a thin metallized glass mirror. This technology is expensive, bulky, and does not scale well to meet the future needs of astronomical and vision science systems. The trend in astronomy is toward telescopes with larger apertures to collect more light and increase the angular resolution. If the telescope is diffraction limited, the resolution limit is given by the Rayleigh criterion:

$$\sin(\theta) = 1.22\frac{\lambda}{D} \qquad (8.1)$$

where θ is the angular resolution, λ is the wavelength of light that is being imaged, and D is the diameter of the aperture that is used for imaging. Currently the largest telescope is at the W.M. Keck Observatory, on the summit of Mauna Kea on the Island of Hawaii; it has an aperture defined by a mirror with a diameter $D = 10$ m [1]. Future telescopes will have apertures that are 30–50 m in diameter [10], [11]. As the telescope aperture increases, the specifications for the deformable mirror that is used to make wavefront corrections become more demanding. Since the portion of the sky that is sampled increases with the telescope aperture, the wavefront error accumulated over the larger aperture also increases. The current piezoelectric deformable mirror technology has approximately 1000 actuators, costs on the order of $1000/actuator, and has a maximum deflection (stroke) of approximately 4 μm. Deformable mirrors for extremely large telescopes will require 10 000 actuators and a stroke of 10 μm because they will be viewing a larger portion of the sky. In addition, multiple deformable mirrors will be required for imaging systems that correct for multiple objects simultaneously (multiobject adaptive optics [MO-AO]) and for multiple heights in the atmosphere (multiconjugate adaptive optics [MC-AO]). The current piezoelectric deformable mirror technology does not scale well in cost and performance to meet these future needs. In addition, increased stroke and decreased cost for deformable mirrors are also required for applications in vision science, which are more cost sensitive and can require up to 15 μm of stroke to correct for aberrations in the eye.

Deformable mirrors based on MEMS technology have been developed to increase the number of actuators and available stroke while decreasing the cost [12], [13]. There are currently MEMS adaptive optic (AO) mirrors with up to 4096 actuators and 6 μm of stroke, although mirrors with both of these specifications are not yet available [14]. In addition, MEMS AO mirrors with up to 10 000 actuators and 10 μm of stroke are in development. Here we consider a MEMS deformable mirror that was initially prototyped in the PolyMUMPS process to fabricate test devices economically and with a reasonable turnaround time, and later commercialized through variations in the standard PolyMUMPS process to meet demanding customer specifications for performance and cost.

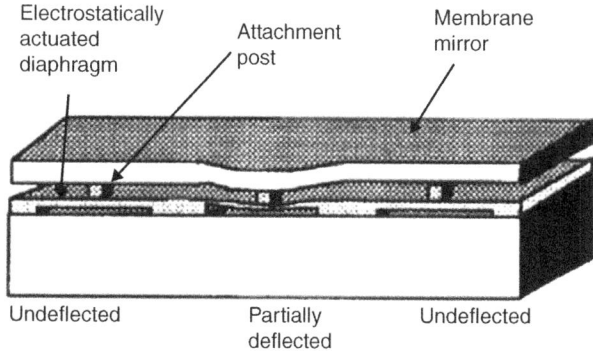

Figure 8.3 MEMS-deformable mirror prototyped in the PolyMUMPS process. (Reprinted with permission from Proceedings of the 5th IEEE International Conference on Emerging Technologies and Factory Automation, *Surface micromachined deformable mirrors*, ©1967 IEEE.) [15].

The initial prototyping was done in 1995 and a company, Boston Micromachines, was started in 1999 to commercialize the MEMS AO technology.

A schematic diagram of the deformable mirror is shown in Figure 8.3 [15]. An electrostatically actuated diaphragm is defined in the 2 μm thick Poly1 layer. The actuator is a fixed-fixed beam that is approximately 350 μm long and 350 μm wide. A deformable mirror membrane is defined in the 1.5 μm thick Poly2 layer and is attached to the actuators by posts defined by POLY1_POLY2_VIA. By applying a voltage between the diaphragm and a counter-electrode defined in the 0.5 μm thick Poly0 layer (not shown), the membrane mirror can be deflected, as shown in Figure 8.4. The maximum stroke is determined by the 2 μm thickness of the sacrificial oxide between Poly0 and Poly1. The actuator "pulls in" or snaps down once the initial 2 μm gap has been decreased by approximately 0.9 μm, providing a controllable 0.8 μm stroke before the pull-in instability is reached. Note that a parallel plate actuator with a linear restoring spring would pull in at 0.67 μm, which is one-third of the initial 2 μm gap. Here the fixed-fixed boundary conditions lead to bending and stretching of the membrane. The bending of the membrane changes the electrostatic force distribution because the middle of the membrane has a smaller gap than the fixed ends of the membrane. The stretching of the membrane stiffens the actuator with increased deflection, which leads to a nonlinear restoring force once the actuator has been displaced by approximately half of the membrane thickness [16].

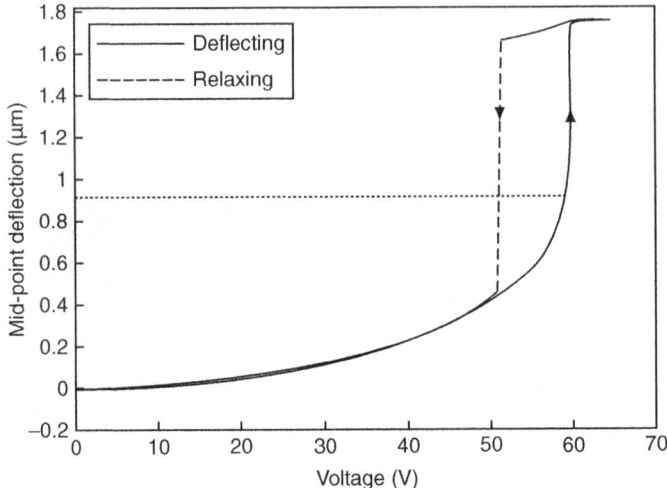

Figure 8.4 Deflection versus voltage for the MEMS mirror. Pull-in occurs at a deflection of 0.9 µm. (Reprinted with permission from Proceedings of the 5th IEEE International Conference on Emerging Technologies and Factory Automation, *Surface micromachined deformable mirrors*, ©1967 IEEE.) [15].

Some of the challenges in the development of a commercial mirror based on this prototype include:

- Topography on the mirror surface. Specifications for some of the more demanding applications call for mirror surfaces that are flat to 1 nm.
- The limited stroke due to the limited thickness of the sacrificial layer, which is 2 µm in the standard MPW process. Some applications call for 10–15 µm of stroke.

To decrease the topography on the mirror surface, a number of approaches were taken. First, the amount of topography that builds up on the surface due to the conformality of the thin-film deposition used in the surface micromachining process was decreased by breaking some of the PolyMUMPS design rules. This approach can be understood by considering the schematic cross section shown in Figure 8.5.

In the figure there is a large gap between two features on the lower layer (dark gray), which results in topography in the upper layer (light gray) equal to the thickness of the lower layer, whereas a small gap results in much smaller topography [17]. The design rules for the PolyMUMPS process call for a minimum separation of 2.0 µm between the features defined in

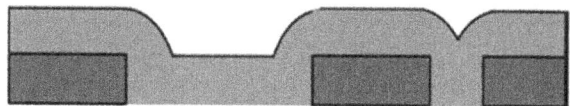

Figure 8.5 Development of surface topography due to conformal coatings. (Used with permission from Raji Krishnamoorthy Mali, Thomas Bifano, and David Koester, *A design-based approach to planarization in multilayer surface micromachining*, J. Micromechanical Microengineering 9, pp. 294–299 [1999].) [17].

Poly1. Since the Poly1 layer thickness is only 2 μm, the layer does not fill in the 2 μm hole that is created when following the design rules. The MEMS designers worked with the foundry processing personnel to find a design strategy to decrease the minimum separation from 2 μm to 1.5 μm. The resulting holes were filled during the conformal deposition as shown in Figure 8.6.

A problem with this approach is that the anchors become weaker as the anchor width decreases. The designers found that they could make a strong anchor by "trapping oxide" in the anchor hole, rather than by removing it. It is possible to trap oxide by not including an etch release hole for the hydrofluoric acid (HF) used during the sacrificial release step. The oxide fills the hole, decreasing the topography due to the hole in the upper layers while preserving the strength of the anchors, as shown in Figure 8.7. In addition to design rule modifications, an additional chemical mechanical polishing step that is not a part of the standard PolyMUMPS process has also been used to smooth the surface topography to nanometer levels of flatness [18].

Another source of topography on the mirror surface resulted from residual stress-gradients in the polysilicon film after it was released, as described previously in Section 4.5 on obtaining flatness in optical MEMS devices. Here it was found that a post-fabrication ion bombardment step could be used to increase the radius of curvature due to residual stress gradients, as shown in Figures 8.8 and 8.9 [19].

The process uses the implantation of inert gases in the mirror surface to create a thin layer of compressively stressed material that acts to flatten the mirror segment. This approach is similar to the stress reduction that occurs due to phosphorous dopant diffusion from the PSG layers during the PolyMUMPS high-temperature anneals, as described in Section 1.2.1 on surface micromachining. In contrast to the anneal step, where dopants diffuse in from both sides, the ion bombardment process only occurs on the top side of the mirror surface,

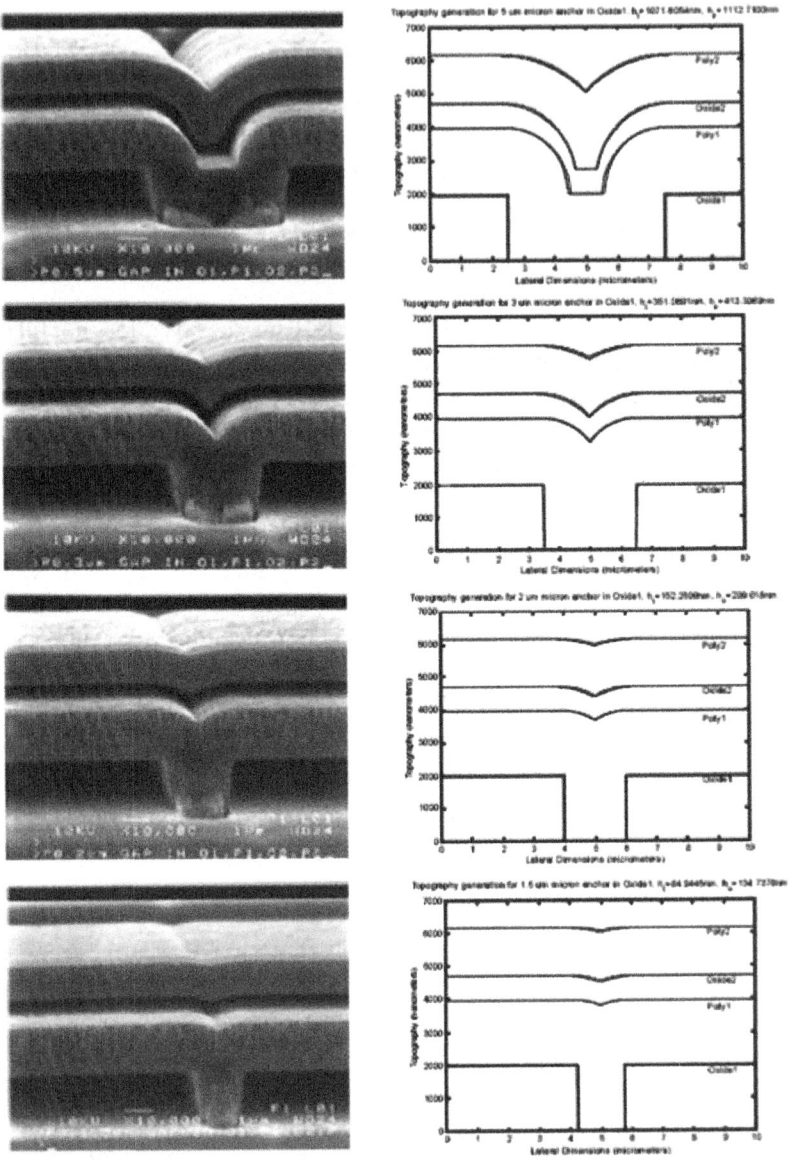

Figure 8.6 Effect of OXIDE1 anchor width on Poly1 and Poly2 topographies with decreasing spacing. Starting from the top figure, the OXIDE1 anchor width decreases from 5 μm, 3 μm, and 2 μm down to 1.5 μm (bottom). (Used with permission from Raji Krishnamoorthy Mali, Thomas Bifano, and David Koester, *A design-based approach to planarization in multilayer surface micromachining*, J. Micromechanical Microengineering 9, pp. 294–299 [1999].) [17].

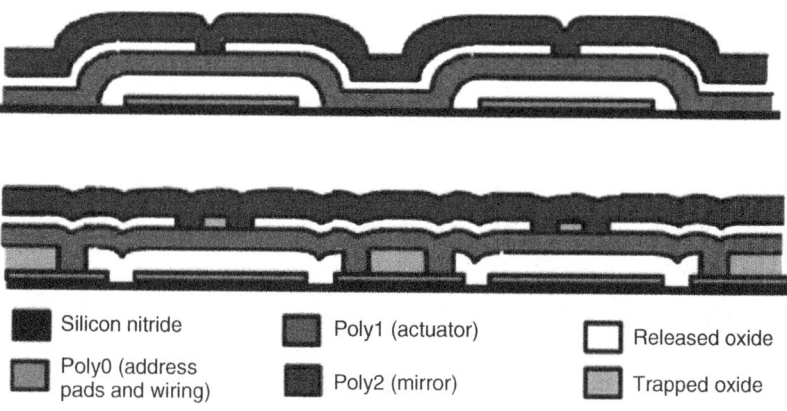

■ Silicon nitride   ■ Poly1 (actuator)   □ Released oxide
▨ Poly0 (address pads and wiring)   ■ Poly2 (mirror)   ▨ Trapped oxide

Figure 8.7 The use of trapped oxide to decrease topography due to conformal coatings in the PolyMUMPS process while preserving the strength of the anchors. (Used with permission from Raji Krishnamoorthy Mali, Thomas Bifano, and David Koester, *A design-based approach to planarization in multilayer surface micromachining*, J. Micromechanical Microengineering 9, pp. 294–299 [1999].) [17]. See color plate section.

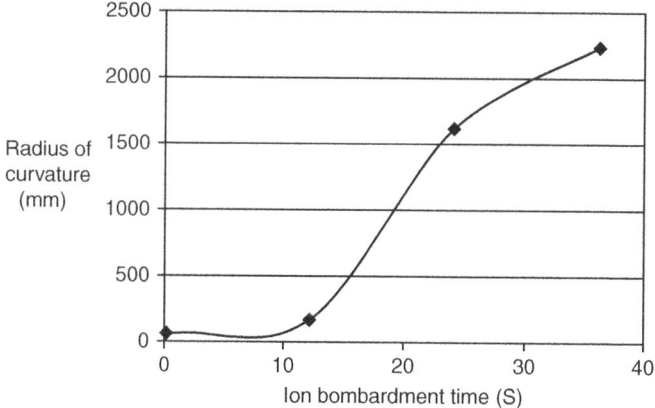

Figure 8.8 Use of ion bombardment to increase the radius of curvature of the mirror. (Reprinted with permission from J. Microelectromechanical Systems, *Elimination of stress-induced curvature in thin-film structures*, ©1967 IEEE.) [19].

so that residual stress gradients can be compensated by creating a counter-acting stressed layer at the surface.

Another source of mirror deformation results from the thin film of metallization that is used to increase the reflectivity of the mirror surface.

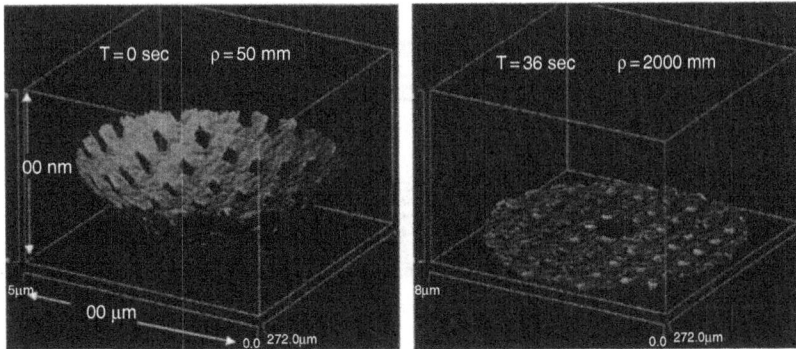

Figure 8.9 Radius of curvature increased from 50 mm to 2000 mm using ion bombardment surface treatment. (Reprinted with permission from J. Microelectromechanical Systems, *Elimination of stress-induced curvature in thin-film structures*, ©1967 IEEE.) [19].

Polysilicon is not very reflective in the visible spectrum that is used in vision science, the reflectivity $R \sim 40\%$ at an optical wavelength of 633 nm, and it is transparent in the IR spectrum typically used in astronomical imaging. The PolyMUMPS process has a gold metallization layer with a thin chrome adhesion layer that is used for making conducting wires and bond pads for probe contacts; however, this layer is 0.52 μm thick and has 50 MPa of residual tensile stress. This level of residual stress can cause significant mirror curvature. While a thick film of gold is desirable for electrical conductivity, a thinner layer of gold, on the order of the optical skin depth (0.12 μm), is all that is required for mirror reflectivity. The designers of the MEMS deformable mirror found that they could apply their own thin metal films with either compressive or tensile stress properties for the mirror's reflective layer that could also be used to counteract the average residual stress-gradient in the polysilicon. Using a combination of ion bombardment and stress compensation from the metallization, the mirrors could be flattened.

The HF etch holes that are included at a 30 μm spacing in the PolyMUMPS process lead to surface topography that can also be detrimental to optical performance. Essentially, the regularly spaced etch holes create a grating that can diffract light. One way to overcome this problem is to eliminate the diffracted light with an aperture stop. Another possibility is to further modify the PolyMUMPS process to enable a sacrificial release etch to be carried out from the back side of the silicon substrate, as shown in Figure 8.10. The through-wafer etch can be either a wet or dry anisotropic etch [20].

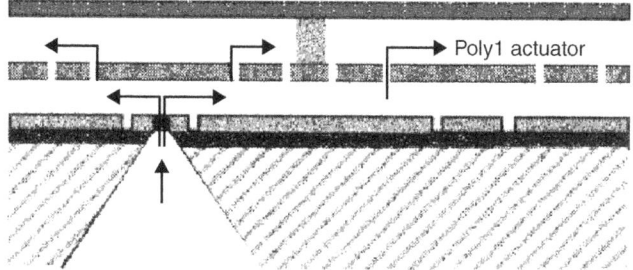

Figure 8.10 Through-wafer etch holes to enable back-side release. The backside etch holes eliminate the need for etch holes in the Poly2 mirror membrane. (Reprinted with permission from IEEE Journal of Selected Topics in Quantum Electronics, *Microelectromechanical deformable mirrors*, ©1967 IEEE.)

Another challenge with using the standard PolyMUMPS process was the limited vertical actuation stroke that could be achieved, as described above. The Poly1 layer was used to fabricate the top membrane for the fixed-fixed beam that formed the top of the actuator, and the Poly0 layer was used for the counter-electrode. The two layers are separated by Oxide1, a 2 μm thick PSG layer. The pull-in instability limits the useful stroke to approximately 0.9 μm; however, several microns of stroke are required for many adaptive optics applications with visible and near-IR light. To increase the stroke, the standard PolyMUMPS process was further modified to increase the thickness of Oxide1 and the use of shaped electrodes (leveraged bending) to enable the use of up to two-thirds of the initial gap. The increase of the sacrificial oxide thickness and the use of leveraged bending enabled a 4 μm stroke to be obtained, as shown in Figures 8.11 and 8.12 [21].

A modified process flow is shown in Figure 8.13. The modifications include:

- A low-stress silicon nitride layer is deposited, lithographically patterned, and etched to allow electrical access to the substrate.
- The sacrificial PSG layer, Oxide1, has a custom thickness that depends on the desired stroke of the device (up to 6 μm).
- The second sacrificial PSG layer, Oxide2, is deposited and chemomechanically polished (CMP) to remove undesired topography resulting from features etched in the underlying layers, greatly improving the surface finish of the final polysilicon mirror layer.

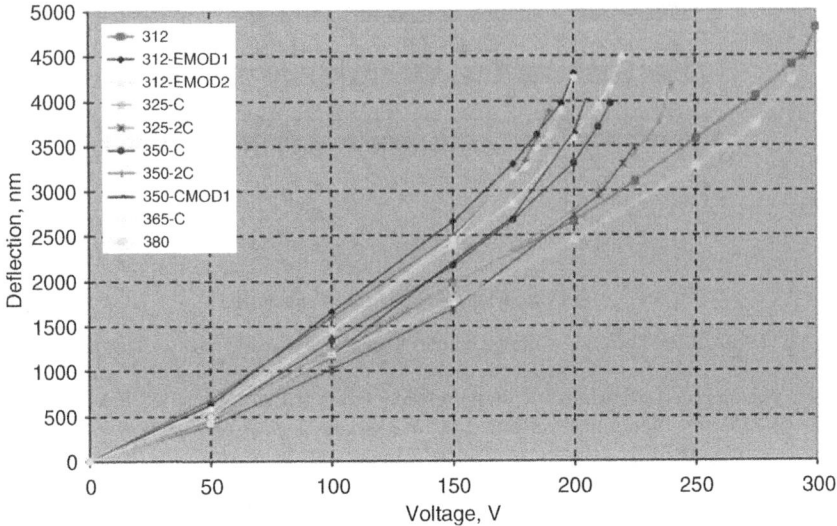

Figure 8.11 Voltage versus deflection characteristics. (Used with permission from S.A. Cornelissen, P.A. Bierden, and T.G. Bifano, *Development of a 4096 element MEMS continuous membrane deformable mirror for high contrast astronomical imaging*, Proc. SPIE 6306, p. 630606–1 [2006].) [21]. See color plate section.

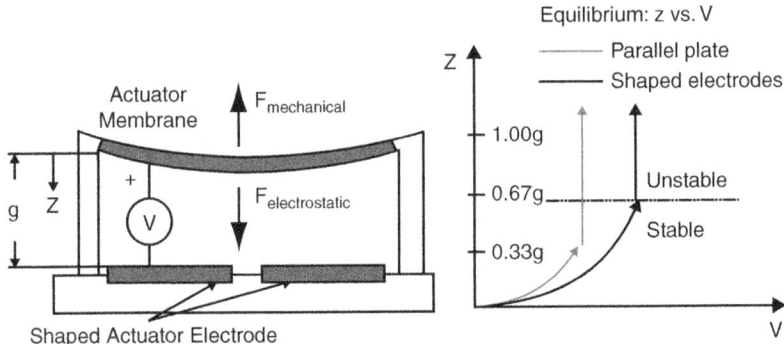

Figure 8.12 Schematic of electrostatic actuation of a double cantilever flexure used in the MEMS DM design. This design uses a shaped actuator electrode to enable increased stroke. (Used with permission from S.A. Cornelissen, P.A. Bierden, and T.G. Bifano, *Development of a 4096 element MEMS continuous membrane deformable mirror for high contrast astronomical imaging*, Proc. SPIE 6306, p. 630606–1 [2006].) [21].

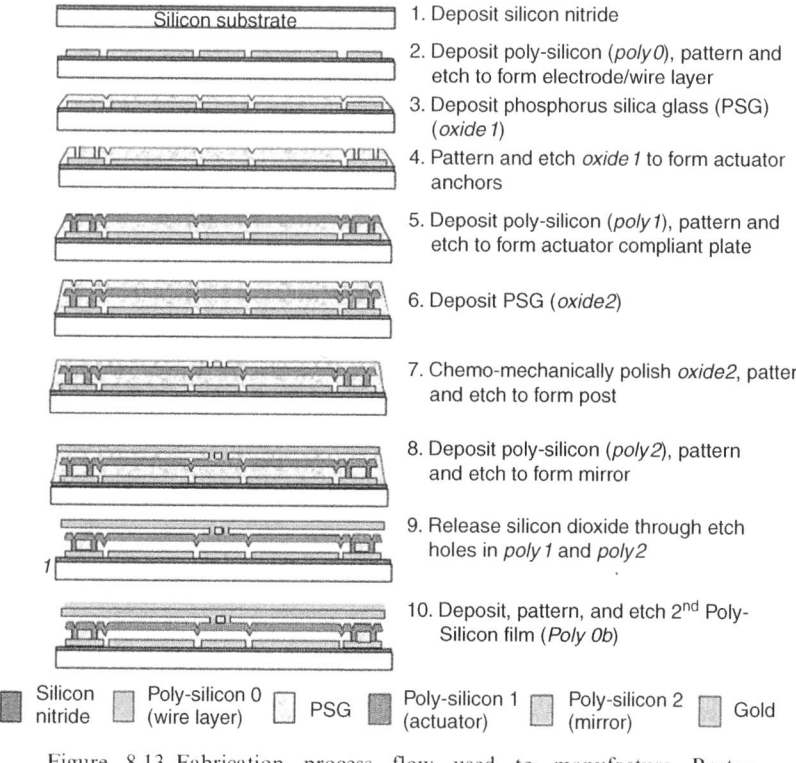

Figure 8.13 Fabrication process flow used to manufacture Boston Micromachines' MEMS DMs. A cross section of a single actuator is shown. (Used with permission from S.A. Cornelissen, P.A. Bierden, and T.G. Bifano, *Development of a 4096 element MEMS continuous membrane deformable mirror for high contrast astronomical imaging*, Proc. SPIE 6306, p. 630606–1 [2006].) [21]. See color plate section.

- The thickness of Oxide2 is increased to provide sufficient clearance for the mirror membrane to enable up to 6 µm range of motion.
- The Poly2 layer is touch polished using CMP to further improve the optical quality of the mirror surface.
- An additional layer of unreleased polysilicon, Poly0a, has been added for wiring below the Poly0b layer that forms the counter-electrode, as shown in Figure 8.14.

Since the initial development started with the robust PolyMUMPS process in a commercial MEMS foundry, the modifications were straightforward and the product development could focus on design variations rather than process development issues.

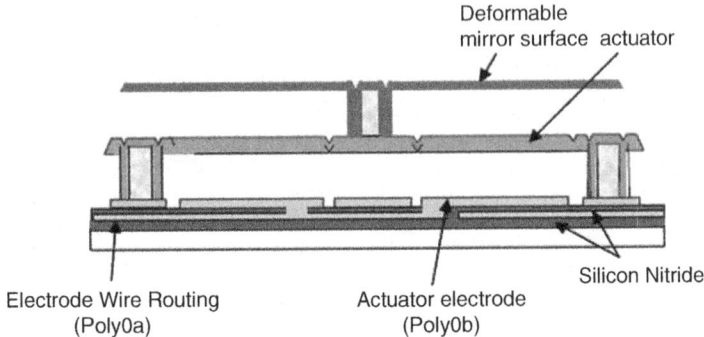

Figure 8.14 Cross section of a single deformable mirror pixel using a buried polysilicon wiring layer to address the actuator electrode. (Used with permission from S.A. Cornelissen, P.A. Bierden, and T.G. Bifano, *Development of a 4096 element MEMS continuous membrane deformable mirror for high contrast astronomical imaging*, Proc. SPIE 6306, p. 630606–1 [2006].)

REFERENCES

1 J.W. Hardy, *Adaptive Optics for Astronomical Telescopes*, Oxford University Press (Oxford Series in Optical and Imaging Sciences, Vol. 16), (1998).
2 R.K. Tyson, *Introduction to Adaptive Optics* (SPIE Tutorial Texts in Optical Engineering Vol. TT41), SPIE (2000).
3 R.K. Tyson, *Adaptive Optics Engineering Handbook* (Optical Science and Engineering), CRC Press (1999).
4 J. Porter, H. Queener, J. Lin, and K. Thorn, *Adaptive Optics for Vision Science: Principles, Practices, Design and Applications* (Wiley Series in Microwave and Optical Engineering), Wiley (2006).
5 M. Booth, *Adaptive optics in microscopy*, Phil. Trans. R. Soc. A 365, pp. 2829–2843 (2007).
6 C. Dainty, *Adaptive Optics for Industry and Medicine: Proceedings of the Sixth International Workshop*, Imperial College Press, London (2007).
7 http://www.ucolick.org/~max/289C/ (last accessed November 12, 2010).
8 C.E. Max, B.A. Macintosh, S.G. Gibbard, D.T. Gavel, H.G. Roe, I. de Pater, A.M. Ghez, D.S. Acton, O. Lai, P. Stomski, and P.L. Wizinowich, *Cloud structures on Neptune observed with Keck Telescope Adaptive Optics*, Astronomical J. 125, pp. 364–375 (2003).
9 Y. Zhang, S. Poonja, and A. Roorda, *MEMS-based adaptive optics scanning laser ophthalmoscopy*, Optics Letters 31, pp. 1268–1270 (2006).
10 J. Nelson and G.H. Sanders, *The status of the thirty meter telescope project*, Proc. SPIE 7012, p. 70121A (2008).
11 R. Gilmozzi and J. Spyromilio, *The European Extremely Large Telescope (E-ELT)*, The Messenger 127, pp. 11–19 (2007).
12 P. Krulevitch, P. Bierden, T. Bifano, E. Carr, C. Dimas, H. Dyson, M. Helmbrecht, P. Kurczynski, R. Muller, S. Olivier, Y.-A. Peter, B. Sadoulet,

O. Solgaard, and E.H. Yang, *MOEMS spatial light modulator development at the Center for Adaptive Optics*, SPIE 4983, pp. 277–234 (2003).

13 N. Devaney, D. Coburn, C. Coleman, J.C. Dainty, E. Dalimier, T. Farrell, D. Lara, D. Mackey, and R. Mackey, *Characterisation of MEMs mirrors for use in atmospheric and ocular wavefront correction*, Proc. SPIE 6888, p. 688802-1 (2008).

14 S.A. Cornelissen, P.A. Bierden, T.G. Bifano, and C.V. Lam, *4096-element continuous face-sheet MEMS deformable mirror for high-contrast imaging*, J. Micro/Nanolith. MEMS MOEMS 8, p. 031308-1 (2009).

15 T.G. Bifano and R. Krishnamoorthy, *Surface micromachined deformable mirrors*, Proc. 5th IEEE International Conference on Emerging Technologies and Factory Automation, Kauai, Hawaii, Nov. 18–21, Vol. 2, pp. 393–399 (1996).

16 E.S. Hung and S.D. Senturia, *Extending the travel range of analog-tuned electrostatic actuators*, J. Microelectromechanical Systems 8, pp. 497–505 (1999).

17 R.K. Mali, T. Bifano, and D. Koester, *A design-based approach to planarization in multilayer surface micromachining*, J. Micromechanical Microengineering 9, pp. 294–299 (1999).

18 J.A. Perreault, P.A. Bierden, M.N. Horenstein, and T.G. Bifano, *Manufacturing of an optical-quality mirror system for adaptive optics*, Proc. SPIE 4493, pp. 13–20 (2002).

19 T.G. Bifano, H.T. Johnson, P. Bierden, and R.K. Mali, *Elimination of stress-induced curvature in thin-film structures*, J. Microelectromechanical Systems 11, pp. 592–597 (2002).

20 T.G. Bifano, J. Perreault, R.K. Mali, and M.N. Horenstein, *Microelectromechanical deformable mirrors*, IEEE J. Selected Topics in Quantum Electronics 5, pp. 83–89 (1999).

21 S.A. Cornelissen, P.A. Bierden, and T.G. Bifano, *Development of a 4096 element MEMS continuous membrane deformable mirror for high contrast astronomical imaging*, Proc. SPIE 6306, p. 630606-1 (2006).

Index

abformung, 17
absorptivity, 99
accelerometer, 34, 49
actuator, 14, 23, 34, 38, 58–61, 63, 65, 68, 70–72,
    77, 79–80, 87, 99, 101–103, 107, 109–110, 112,
    115, 117, 128, 130–131, 133, 134–135, 138,
    141, 142, 146–147, 153, 154
adaptive optics, 13, 32, 95, 144, 145, 153,
    157
adhesion, 12, 152
air-bag, 50
alignment, 11, 14, 21
all angle, 27
aluminum, 1, 78
analogy, 45, 98
analytical, 5, 68–69, 77, 85
anchor, 9, 11, 19, 21–23, 26, 28, 30, 40, 68,
    70, 84, 103, 109, 136, 139, 149, 150
angular resolution, 146
anisotropic etching, 17
anneal, 6, 7, 10–11, 14, 19, 114, 149
aperture stop, 152
aquadag, 140
arbitrary waveform generators, 139
aspect ratio, 14, 17
assembly, 82, 84
astronomy, 93, 144, 145
attenuator, 14
AutoCAD, 23
auxetic, 39
axial strain, 38
axial stress, 40, 55

back-bending, 115
backend, 2
backside contact, 140

back-side release, 153
band-pass, 87
battery, 59, 79
battleground, 136
bearing, 28
bending, 35, 41, 46, 111, 133, 147, 153
bent-beam, 47, 107, 109, 137
bimorph, 12
black-body, 100
blood sugar level, 118
bolometer, 113
bond pads, 9–10, 12, 25, 28, 53, 83, 103, 107,
    136–139, 152
bonding, 6, 21, 77, 135
Boolean, 26
Boston Micromachines, 147, 155
boundary condition, 41–42, 63, 93, 105, 106,
    110, 130, 147
boxes, 27
Bragg dielectric stack, 87–88, 90
breakdown, 95
buckle, 37, 46, 93, 116–117
bulk micromachining, 14, 17, 32, 135–136
buoyant force, 100
buried oxide, 14, 15, 16

Caltech Intermediate Format (CIF), 5, 24
cantilever, 3, 29–30, 41–42, 49, 50, 55,
    67–70, 74–76, 110
capacitance, 58–61, 70, 72, 79,
    112–113
capacitor, 58–60, 68, 72
capillary, 7, 12, 120, 124, 126, 128, 134, 136
cells, 24, 28–29
channel, 3, 82, 134
charge, 7, 58–59, 78–79, 102, 140

Index

checksum, 24
chemical mechanical polishing (CMP), 13, 95, 149, 153, 155
chrome, 12, 152
circles, 27
clamped circular diaphragm, 45, 52, 128
clamped square membrane, 52–53
clamped-clamped beam, 104
coefficient of thermal expansion (CTE), 3, 93, 102, 106, 110, 116
cold arm, 103–104, 107, 115–116
collimator, 82
colloidal suspension, 140
color palette, 27
comb-drive, 28, 53–55, 58, 60–61, 71, 73, 140
command line, 25
commercialized, 146
complementary metal oxide semiconductor (CMOS), 1, 24, 88, 91
compressive stress, 7, 10–11, 37, 46, 47, 93, 116, 137
compressor, 49
computer aided design (CAD), 5, 23
conducting glue, 140
conduction, 98, 115
conductivity, 18, 98, 101, 113, 115, 117, 140, 152
conformal coating, 9, 94, 149
consequences, 23
constrained, 8, 93, 102–103
contact angle, 125
contact resistance, 140
contaminant, 139
convection, 98, 100
conversion factor, 25
Corning 7740, 77–78
Couette flow, 121
counter-electrode, 7, 62, 67, 77–79, 86, 94, 129, 130
coupled domain analysis, 5
Coventor, 23
crash sensor, 51
creep, 83
critical point drying (CPD), 12, 135
critical stress, 37
critical voltage, 62
cross-connect optical switch, 74
cross-section, 1, 2, 7, 13, 18, 30, 31, 44, 53, 55–56, 90, 128, 131, 148, 155

curvature, 12, 32, 84, 93, 111, 149, 151–152
curved features, 24
curves, 27

DALSA, 24
dampening, 48, 49
decreased cost, 146
deep reactive ion etching (DRIE), 16
deflection angle, 77
deformable mirror, 144–148, 152, 154
deformations, 10, 14, 34, 77, 93, 106–107, 115, 120, 136, 145
density, 54–55, 78, 98, 100, 106, 118, 127, 133
deposition, 1–3, 8, 9, 14, 16, 18, 93, 148–149
desiccant, 139
design rule checking (DRC), 24–25
design rules, 6, 11, 13, 21, 23, 61, 148
design variations, 155
designer, 21, 136
device layer, 14, 15, 16, 55–56, 77, 88, 93
diaphragm, 29–30, 45, 52, 128, 131, 133, 147
dicing, 135
die level, 5, 135–136
dielectric, 29, 59, 68, 87–88, 90–91, 95, 131, 133
diffraction, 74, 146, 152
diffusion, 5, 14, 115, 123, 149
digital light processing (DLP), 43
digital projectors, 74
dimple, 7, 8, 12, 133
displacement, 41, 53, 62, 65, 70, 72, 103, 106, 108, 128, 133–134, 141, 143
display, 26, 43, 45, 74–75, 95
distributed load, 42–43, 49, 52
divergence, 118
dopants, 1, 149
doping, 5, 7, 10, 89
double-sided sticky tape, 137, 140
dovetails, 82–83
drawing exchange format (DXF), 24
drawing toolbar, 27–28
driving force, 99
drop ejector, 127–128, 129, 131
dry environment, 139
dynamic mirror deformations, 77
dynamic pressure, 128

eddy, 120
EFAB, 25
elasticity, 34, 44, 111
electric field, 78
electrical cross-over, 19
electrical test structures (E-Test), 45, 139
electrodeposited, 17
electrokinetic flow, 119
electrokinetic pump, 134
electrolysis, 129, 131
electromechanical filter, 3
electromechanical modeling, 76, 80
electro-osmosis, 134
electrophoresis, 134
electroplating, 6, 17, 19, 25
electrostatic actuation, 23, 58–59, 61, 63, 65, 68, 72, 77, 79, 102, 127–128, 131, 138–139, 141, 154
electrostatic force, 23, 34, 60–61, 71–72, 75, 91, 130, 133, 140, 147
electrostatic motor, 11
ellipses, 27
emissivity, 99
encapsulation, 18
enclose, 22
etch holes, 7, 11, 152–153
etch stop, 14, 16, 21
etching, 1, 2, 12, 18, 21, 83, 135
Euler buckling, 37, 116, 137
eutectic, 77, 114–115
exclusion zone, 16
extension strain, 38
externally referenced cell, 24

Fabry–Perot interferometer, 56, 57, 68, 86–89, 92
fan-out, 139
fatigue, 83
feedback, 72
fidelity, 21, 140
file transfer protocol (FTP), 24
filter, 3, 87–89, 91, 95
finesse, 87, 91
finite element analysis (FEM), 55
fixed-fixed beam, 42–43, 45, 46, 47, 55, 68, 70, 115–116, 137, 147, 153
fixed-free beam, 11, 41–43, 55, 67–68, 111
fixed-guided beam, 42, 50, 54–56, 63, 65–66
flat mirror, 84

flatness, 88, 91, 149
flexural rigidity, 52
flexure, 103–104, 115–116, 130, 154
floaters, 10
fluidic elements, 28
folded spring, 28, 53–54, 65, 91
fold-up mirror, 56
footing, 16
foundry, 6, 13, 24, 149, 155
Fourier's law, 98, 101
friction, 48, 128
fringing, 70
fuzzy blob, 144–145

galvanoformung, 17
galvo scanner, 74–75, 77
ganged, 108
gap, 23, 48, 53, 58–59, 60–61, 63, 65, 68–69, 72, 75, 77–78, 82, 84, 87–91, 122, 129–130, 147–148, 153
gate electrode, 3
gimbal, 82–83, 86
glass wafer, 77–78
gold, 11, 12, 14, 16, 18, 21, 76, 93, 114, 115, 152
gradient, 3, 5, 10, 45, 46, 70, 78, 84, 88, 93, 98, 101, 106, 114, 120, 136–137, 149, 151–152
graphic data system (GDSII), 5, 24
graphical solution, 62
grating, 13, 152
gravitational constant, 127
gravitational potential energy, 127
grey-body, 100
grid, 25–26
ground pad, 140
ground plane, 7, 29, 45, 68
grow, 26
guide star, 144–145

handle wafer, 15
hard mask, 10
HARM, 24
HBSRI, 25
heat balance equation, 101
heat capacity, 99–101
heat energy, 100
heat flow, 98–100,
heat flux, 100
heat transfer, 98–100, 106

Index

heater, 114, 117, 127
heatuator, 103–104, 106, 108, 115
hierarchical, 24
high-magnification objective, 141
high-temperature anneals, 149
high-Q, 3, 141
high-resistivity, 18
high-stress, 84
high-volume manufacturing, 6
hinge, 11, 31, 83
hinge pin, 31
hinge staple, 31
holding force, 84
hole, 26
Hooke's law, 34–35, 61
hot arm, 103–104, 106–107, 115–116
hub, 8, 11, 31
hydrofluoric acid, 7, 12, 16, 21, 136, 149, 152
hydrophilic, 125–126
hydrophobic, 126
hyperbola, 27

ice cream maker, 121
image aberrations, 144, 145
iMEMS, 25
implantable medical devices, 118
incompressible fluid, 118
index of refraction, 87, 144
inertia, 36, 44, 70, 112, 119, 121
infrared, 113, 145
ink, 114, 117, 127–129, 131, 133
inkjet, 114, 117–118, 124, 127–128, 131, 133
instability, 45, 63, 65, 68, 147, 153
instances, 27
insulation, 18
insulin, 118
INTEGRAM, 24
integrated circuit, 5
IntelliMask, 23
IntelliSense, 23, 51, 76, 80
IntelliSuite, 51, 76, 80, 107
interface, 7, 99, 125–126, 135–136
interface energy, 125
interface force, 125
interference, 74, 130, 139, 141
internal data format, 24
internal unit, 25
interposer, 139
ion bombardment, 149, 151–152
ion implantation, 19
IR radiation, 89

IR spectrometer, 91
isolation, 18
isolation oxide, 21

Jale3D, 23
Joule heating, 100–102, 103, 106, 115

kerf, 135
kinetic energy, 128
known good designs, 29
KOH potassium hydroxide, 17, 21, 136

lab-on-a-chip, 118–119
laminar, 119, 123
landing a probe, 137
landing pads, 83, 86
laser communications, 144
layer palette, 25
layer properties, 26
layer setup, 26
layout, 5, 23–26, 28–29, 31, 45, 53–55, 56, 61, 86, 135, 139, 140
L-Edit, 23, 25, 28–29, 53, 55, 68
leveraged bending, 153
libraries, 25
library elements, 28
lift-off, 14
LIGA, 17
line scan, 143
linear comb-drive, 28
location information, 25
locator bar, 26
locator units, 26
locking mechanism, 17
low pressure chemical vapor deposition (LPCVD), 7
low-stress nitride, 18, 19
lucent lambda router, 82
lumped parameter model, 99

magnetic, 17
major grid, 26
mandatory rules, 23
mask, 1, 2, 7, 11, 14, 16, 18–19, 21–24
mask levels, 23
mass, 48–51, 65–67, 74, 112, 115, 118, 135
materials properties, 45, 70

Mauna Kea, 146
MCNC, 6
measuring distances, 26
mechanical force, 31, 138
mechanical linkage, 18
mechanical probing, 138–139
mechanical properties, 3, 14, 45, 68
mechanical work, 79
melting, 114–115
membrane, 41, 52–53, 84, 93, 123, 128, 129, 130, 133, 147, 153, 155
memory, 83
MEMS design, 6, 28, 41, 68
MEMS fabrication process, 1, 3, 21
MEMS Pro, 23, 25, 28–29, 53, 55, 69
MEMSCAP, 7, 8, 15, 19, 24
MEMSJet, 129, 131
meniscus, 12, 124, 134, 136
menu, 25
merge, 26
metal, 14, 18–19
metallization, 12, 14, 16, 78, 83–84, 86, 93–94, 151
MetalMUMPS, 17–18, 20, 24
microelectromechanical systems (MEMS), 1, 34
MicroFabrica, 25
micro-fluidic, 17, 118–119, 121
micromanipulators, 138
micromechanical, 13, 32, 135
microprobe, 17
microrelay, 18
microscope, 2, 45, 137–138, 141
mid-wavelength IR, 88, 89
minimum design rules, 23
minimum feature size, 17, 21, 65
minor grid, 26
mirror, 12–14, 17, 32, 43, 56, 68, 77–84, 85, 86–87, 89, 91, 93–95, 144–149, 152–153
mirror tilt angle, 77
mix, 119
modeling, 25
modulus of elasticity, 34, 44, 111
moment of inertia, 36, 44, 70
motors, 28, 32
mouse pointer, 26
MPK, 24
M-Test, 29, 30, 45–46, 67–69, 73, 128, 140

multi-conjugate adaptive optics, 146
multi-meter, 139
multi-object adaptive optics, 146
multiple probe contacts, 139
multi-project wafer (MPW), 6, 24

Neptune, 144, 145
Newton's law, 99, 118
Newton's rings, 130–131
Newtonian fluid, 120
nickel, 17, 19
night vision, 113
nitride, 1, 7, 9, 11, 17–18, 19, 21–22, 68, 89, 91, 133, 153
nominal design rules, 23
non-linear spring, 62, 65, 72, 130
non-Manhattan geometries, 24
non-sacrificial, 7
novices, 23
nozzle, 114, 124, 127–129, 134

oil dashpot, 48
optical cavity, 86–87
optical cross-connect switch, 82
optical fiber, 82, 84
optical MEMS, 17, 74, 93, 149
optical quality, 14, 155
optical surface, 93
optical switch, 13, 81–82
optimize, 82, 91
orthogonal, 27, 47
output data files, 24
over-etching, 21
oxide, 1, 6, 7, 9, 11–13, 16–19, 21, 30, 63, 83, 88, 90–91, 114, 129, 133, 147, 149, 151, 153

packaging, 128, 135
parabola, 27
parallel plate actuator, 58–60, 63, 65, 68, 70, 72, 80, 134, 147
parallel plate capacitor, 58, 68
parallelism, 88
partially reflecting mirrors, 86
path-to-the-sea, 6
pattern transfer fidelity, 21
peristaltic pump, 134
permittivity of free space, 59, 134
phase boundary, 136

phosphorous, 115, 149
phospho-silicate glass (PSG), 7
photodiode, 88
photolithography, 2, 13, 21
photoresist, 1, 2, 19
photosensitive epoxy, 129
pie wedges, 27
piezoelectric actuators, 128
piezoresistive sensor, 138
piezoresistor, 52
piston drop ejector, 131–132
planarize, 94, 95
Planck's constant, 114
plastic deformation, 115
platform, 112–113
plating base, 19, 21
point force, 35, 45, 56, 75
point load, 41–42, 55, 67, 68, 74
Poiseuille flow, 122
Poisson's ratio, 38–39, 41, 56, 68, 70
poke, 139
polishing, 13, 94–95, 149
polygons, 27
polyimide, 78, 88
polymer, 1, 16, 39, 129
PolyMUMPS, 7, 8, 10–11, 13–14, 21, 24, 26, 29, 31–32, 51–53, 63–65, 67, 82–84, 88, 93–94, 112–113, 117, 128, 133, 146–149, 151, 152–153, 155
polysilicon, 1, 2, 6, 7, 9, 11, 13–14, 17–18, 21–22, 34–36, 38–39, 41, 45, 47, 50, 52, 53–54, 63, 66, 68, 73–74, 77, 83–84, 91, 93, 113–114, 116–117, 136, 149, 152, 155
ports, 27, 137
post-processing, 24, 135
potential, 59, 78–79, 86, 98, 119, 127, 129, 140
potential energy, 59, 79
power balance equation, 100–101
PPK, 24
precision, 25
pressure, 44, 52–53, 119, 121–122, 124, 128, 130, 131, 133–134, 136
print through, 94
printing, 118, 124, 127, 128, 133–134
probe station, 137–138, 140
probe-card, 139
probes, 137–139
process development, 3, 6, 11, 155
process latitude, 21

process parameters, 82
process setup files, 23
product manufacturing, 6
productivity, 29
programmable function generators, 139
projection display, 74–75, 95
proof mass, 50
proof of concept, 6
propellant, 114
protective coating, 16
prototyping, 5, 6, 13, 24, 33, 74, 93, 147
PSG, 6–7, 9–11, 14, 17–19, 115, 149, 153
pull-in, 45, 57, 62–65, 68–70, 72, 76, 80, 85, 130, 140, 147, 153
pump, 99, 134
Pythagorean theorem, 108

QinetiQ, 24

radiation, 98–100
radius of curvature, 84, 93, 149, 151
Rayleigh criterion, 146
Rayleigh Ritz energy method, 55
reactive ion etching (RIE), 16, 22
reflection, 74, 130
reflectivity, 77, 86, 90, 91, 93, 99, 151
refraction, 74, 87–88, 144
relay switch, 18
release, 2, 3, 5, 7, 8, 10–12, 21, 26, 29, 84, 93, 135–136, 139, 149, 152
reliability, 83
residual stress, 3, 5, 10, 45, 46, 47, 68, 84, 93, 108, 136–137, 149, 151–152
resistance, 7, 10, 11, 19, 98, 100, 104, 106, 107, 115, 120, 138, 140
resistivity, 18, 26, 45, 98
resonance, 65, 67, 140
resonant gate transistor, 1, 3, 32
resonator, 28, 53–55, 66–67, 140
response, 34, 40–41, 113, 120
restoring force, 43, 49, 61–62, 65, 72, 147
retina, 144, 145
Reynolds number, 121, 123
RF, 17, 131
Richard Feynman, 31
Roark's formulas for stress and strain, 41, 57

rods and cones, 145
rotary comb-drive, 28
rotary motor, 28
rotary pump, 134
rotary torsional springs, 28
rotor, 11
roughness, 93, 141
rulers, 27

sacrificial etch, 2, 9–10, 12, 21, 23, 28
sacrificial layer, 9, 19, 29, 63, 135, 148
sacrificial oxide, 8, 12, 13, 18, 21, 65, 83, 90, 129, 147, 153
sacrificial release, 10, 12, 30, 135, 149, 152
Sandia Ultra-planar Multi-level MEMS Technology, 13
scaling, 31, 54, 65, 119
scanner, 77, 85, 95
scanning electron microscope, 140
sealing, 135
segmented mirror, 95
selectivity, 21–22
self assembly, 82
self-assembled monolayer (SAM), 136
semiconductor fabrication process, 1, 2
sensor, 47, 50, 52, 107, 112–113, 118, 136, 138, 145
serpentine, 83, 113
setup design, 26
setup layers, 26
shadow mask, 16
shaped electrodes, 153
shapes, 24, 27–28
shear modulus, 40, 44, 80, 120
shear rate, 120
shear strain, 40, 120
shear stress, 40, 120
shearing force, 40
short working distance objective, 141
short-wavelength IR, 88
shrink, 26, 39
shuttle plates, 28
sidewalls, 18, 21, 82–83
silicon nitride, 1, 7, 9, 133, 153
silicon on insulator (SOI), 14, 15, 16, 24, 33, 55, 56, 77, 88, 93
silicon wafer, 18, 77–78
single crystal silicon, 14, 34, 39, 74
sinusoids, 27

skin depth, 152
skin on a drum, 93
sluggish, 139
Snook's law, 72
SoftMEMS, 25, 95
SOIMUMPS, 14, 15, 24, 55, 74, 77
solid model, 24, 30, 54–55, 69, 85
solid source diffusion, 14
source-drain channel, 3
spacer layer, 9, 77–78
special curves, 27
spirals, 27
splines, 27
spring, 28, 32, 34, 35–36, 37, 41, 48–50, 54–55, 61–63, 65, 67–68, 70, 72, 83, 85, 91, 139, 147
spring constant, 34–37, 49–50, 54–55, 62–63, 65, 67–68, 70, 80
springs connected in parallel, 36
springs connected in series, 36–37, 54, 65, 72, 91
squeezed film dampening, 48
stacking, 12, 65, 84–85
standard cells, 28–29
standard components, 25, 28
staple, 11
static mirror deformations, 77
stencil, 19, 21
Stephan-Boltzmann law, 99
stiction, 8, 12, 16, 136–137
stiffness in shear, 120
stimulus, 34, 40, 120
Stoney's formula, 111
stop, 16
strain, 34, 38–40, 44, 47, 52, 102, 107, 111, 120, 137
stress gradient, 6, 10, 93
stress-induced curvature, 12
stress parameter S, 46
stress relief, 83
stringer, 23
stroboscopic image, 130
stroke, 95, 146–148, 153–154
structural, 7, 8, 9–11, 18, 26, 29, 63
SU-8, 129
sub-dicing, 135
substrate, 3, 6, 7, 9–11, 14, 16, 18, 29, 30, 46, 49, 60, 86, 88–89, 91, 103–105, 106, 109–112, 115–117, 135–136, 139, 140, 143, 152–153
SUMMiT, 13, 25, 94–95, 131–132

Index

supercritical drying process, 136
superheated, 114
surface energy, 123, 125, 128
surface finish, 153
surface heat transfer coefficient, 99
surface micromachining, 2, 6, 7, 8, 9, 13–14, 17, 63, 65, 112, 114, 128, 132, 135–136, 148, 149
surface roughness, 93, 141
surface tension, 119, 123–125
surface treatment, 152
surround, 11
suspension, 112
swimming, 119, 123
switch, 18, 74, 81–82, 84, 138

technology, 25
technology units, 25
teeth, 17, 60–61
Teflon-like coating, 136
temperature distribution, 104, 111
temperature gradient, 98, 106
tensile stress, 7, 38, 47, 93–94, 137, 152
test structures, 25, 28, 30, 33, 45, 46, 57, 68–69, 136, 139–140
thermal actuator, 14, 18, 38, 102, 107, 108, 114–115, 117
thermal bimorph actuator, 109, 115
thermal capacitance, 112–113
thermal conductivity, 98–99, 101, 113, 115, 117
thermal current, 98
thermal electrical, 106
thermal equilibrium, 101
thermal expansion, 101–102, 104, 106, 109, 111–112
thermal mass, 112, 115
thermal oxide, 18
thermal path, 109
thermal potential, 98
thermal resistance, 98
thermal stress analysis, 106
thermal time constant, 112–113
thermally isolated, 112–113, 136
thermomechanical analysis, 106
thin film properties, 140
thin-film deposition, 1, 2, 3
thin-film parameters, 6
three-dimensional optical switching, 81
through-wafer etch, 16, 88
through-wafer holes, 16

tip-tilt mirror, 43, 77, 84, 87
toolbar, 27
topography, 2, 9, 13, 16, 21, 23, 86, 94, 148–149, 152–153
torque, 43–44, 79, 80
torsion, 17, 43–44, 56, 80, 83
torsion rod, 43–44, 56, 80
torsional mirror, 77, 78, 80, 86
torsional moment of inertia, 44
torsional spring, 43, 80, 83
toruses, 27
trampoline, 123
transmission coefficient, 89, 90
transmission function, 92
transmission peaks, 91
transverse strain, 38
trapped oxide, 91, 149, 151
Tronics Microsystems, 24
trusses, 54
tunable Fabry-Perot optical filter, 95
tunable lasers, 74
turbulent flow, 123
two-dimensional optical switching, 81
two-axis torsional mirror, 86

undercutting, 16, 18

vacuum hold down, 137
Van der Waals attraction, 7, 12
vapor bubble, 114, 117, 127
vapor phase, 16
variable optical attenuator (VOA), 14, 74
variable voltage power supply, 139
velocity dependent force, 48
velocity gradient, 120
vernier, 47
vias, 11
violate a design rule, 23
virtual work, 59, 79
viscometer, 122
viscosity, 119–121
viscous flow, 121
viscous forces, 121, 123
viscous loss, 48
viscous pressure drop, 128
vision science, 144–146, 152
volumetric flow rate, 134
vortices, 123

W.M. Keck Observatory, 146
wafer bonding, 6

wafer level layout, 135
walk on water, 119, 123
wavefront, 32, 144–146
wavefront sensor, 144
wheel, 8, 31
white light non-contact interferometer, 141
window, 74
wire bonding, 21, 135
wires, 12, 18, 23, 27, 86, 94, 113, 115, 139, 140, 152, 155–156
witness mark, 139

X-beam electrostatic actuator, 63–65, 141, 142

yield, 82–83
Young's equation, 125
Young's modulus, 34, 41, 45, 46, 54–56, 64, 68, 102, 111, 116, 133

z-beam electrostatic actuator, 63–65
z-height data, 141

Made in the USA
Monee, IL
03 May 2026